NETS, PUZZLES, AND POSTMEN

NETS, PUZZLES, AND POSTMEN

Peter M. Higgins

OXFORD
UNIVERSITY PRESS

OXFORD

UNIVERSITY PRESS

Great Clarendon Street, Oxford OX2 6DP

Oxford University Press is a department of the University of Oxford.
It furthers the University's objective of excellence in research, scholarship,
and education by publishing worldwide in

Oxford New York

Auckland Cape Town Dar es Salaam Hong Kong Karachi
Kuala Lumpur Madrid Melbourne Mexico City Nairobi
New Delhi Shanghai Taipei Toronto

With offices in

Argentina Austria Brazil Chile Czech Republic France Greece
Guatemala Hungary Italy Japan Poland Portugal Singapore
South Korea Switzerland Thailand Turkey Ukraine Vietnam

Oxford is a registered trademark of Oxford University Press
in the UK and in certain other countries

Published in the United States
by Oxford University Press Inc., New York

British Library Cataloguing in Publication Data
Data available

Library of Congress Cataloging in Publication Data
Data available

Typeset by SPI Publisher Services, Pondicherry, India
Printed in Great Britain
on acid-free paper by
Biddles Ltd., King's Lynn, Norfolk

ISBN 978–0–19–921842–4

1 3 5 7 9 10 8 6 4 2

PREFACE

This is the third book of this type that I have written for OUP that celebrates the everlasting wonder of mathematics. The first, *Mathematics for the Curious* (1997) was general in nature while *Mathematics for the Imagination* (2002) concentrated more on geometry. The topic of networks was touched on in both those books but it deserves a fuller treatment in its own right.

There are several reasons for this. From a mathematical viewpoint, networks have come of age as they have invaded one branch of the subject after another. What is more, networks themselves are all around us from age-old examples such as family trees to the modern phenomenon of the Internet and World Wide Web. On the other hand networks can be introduced to anyone and, as with sudoku puzzles, they can immediately begin playing with them as 'they don't involve any mathematics, just logic'. This description has brought a smile to many a mathematician's face as they know better. At the same time, however, they do appreciate what people mean when they express these sentiments. It would be better to say that the topic does not, at least at first glance, require all the mathematics we might have heard about in school such as arithmetic and algebra, geometry and trigonometry, and so on.

The text is designed to be read straight through. We go from simple examples and ideas to a host of applications from various games and puzzles, including sudoku, onwards to a variety of topics, some serious, such as planning routes and maximizing profits, and some seemingly not so. As often happens, mathematics is indifferent to the applications we have in mind when we introduce a type of problem and it can come to pass that 'serious' and 'recreational' problems turn out to be related or in some cases one and the same.

There is no denying the existence and importance of networks in the world for they underpin the nature of traffic flow, both in communications and in the movement of solid vehicles. Although these topics are discussed, I want to continue the style of my previous books and let the subject matter percolate up through a variety of simple examples to engage the reader. That is why you will meet so many different problem types from games to automata, postal routes to map colouring, matchings to RNA reconstruction. The world of networks is a wide one.

The descriptions by and large do not assume mathematical knowledge or habits of thinking and the development is based around explaining what we mean by certain ordinary words in particular contexts and straightforward 'logical' argument. However, I know some of my readers do know a thing or two about mathematics and would sometimes appreciate more explanation. For those of you who wish to pursue these things without having to chase up another source, the final chapter, 'For Connoisseurs', does give more in the way of mathematical explanation. An asterisk in the text tells the reader that more is said on a particular matter in this final chapter.

Peter Higgins
Colchester (2007)

CONTENTS

1. Nets, Trees, and Lies   1

Trees   5
Chemical isomers   9
Lying liars and the lies they tell   10

2. Trees and Games of Logic   17

Familiar logic games   17
Exotic squares and Sudoku   23

3. The Nature of Nets   35

The small world phenomenon   35
The bridges of Königsberg   43
Hand-shaking and its consequences   48
Cycles that take you on a tour   53
Party problems   56

4. Colouring and Planarity   63

The four-colour map problem   63
How edges can ruin planarity   74
Rabbits out of hats   80
    1. Guarding the gallery   81
    2. Innocent questions of points and lines   84
    3. Brouwer's fixed point theorem   90

5. How to Traverse a Network   101

The Euler–Fleury method   101
The Chinese Postman Problem   105

6. One-Way Systems 111

Nets that remember where you have been 114
Nets as machines 119
Automata with something to say 128
Lattices 132

7. Spanning Networks 137

Sorting the traffic 140
Greedy salesmen 145
Finding the quick route 147
The P versus NP controversy 150

8. Going with the Flow 159

Network capacities and finding suitable boys 159
Marriage and other problems 163
Harems, maximum flows, and other things 168

9. Novel Applications of Nets 175

Instant Insanity 175
Sharing the wine 181
Jealousy problems 184
Mazes and labyrinths 185
Trees and codes 188
Reassembling RNA chains 191

10. For Connoisseurs 197

References 237

Further Reading 239

Index 243

1

Nets, Trees, and Lies

The importance of networks has taken everyone by surprise. So much of modern mathematics is about how one thing is related to another, or more widely, how objects within a collection are interrelated, and this idea is captured in the notion of a network. And networks are just what you would imagine them to be—they can be pictured as an array of dots on the page, called *nodes* or *vertices*, with some of the dots joined to others by lines that we call *edges* or *arcs*. These links could stand for physical connections by bridges, roads, or wires, or less tangible connections by radio signals, or abstract personal connections such as friendship or enmity, or even the ancestral relationships of a family tree. This book reveals something of the surprising and subtle nature of networks, or *nets* as they can sometimes be called. Like nets themselves, it does not have a linear structure but, as you read on, the overall picture will become progressively clearer and the many sides of the subject will begin to coalesce.

Who discovered networks? The question is almost like asking who discovered drawing—the urge to start doodling pictures of networks is almost overwhelming as soon as we begin thinking about a situation in which there is a multitude of connections. The advantage of the picture is that it allows you to see all the connections at once and we can remind ourselves of any one of them simply by flitting our eyes around the diagram.

Perhaps networks have been underestimated because they are so common, yet at the same time they seem to lack any structure. The mathematical topics that have been studied extensively for thousands of years are numbers and geometry. Numbers are pervasive, they allow us to tally and compare, and have an undeniable natural order. Geometric objects are pretty and visual, providing all manner of symmetries that can strike you before a word is said, so the attraction of geometry is very powerful and immediate. Networks on the other hand are none of these things. Networks are not numbers of any kind, nor are they truly geometrical even though we can draw pictures of them. They represent quite a different realm of mathematics. And not only of mathematics, for everyone appreciates the importance of networking—the real measure of our comprehension of the world is our understanding of how all the various parts come together and affect one another.

Moreover, the use of the word 'network' in this context is more than just a metaphor. Some of the most difficult and technically demanding research in the social and political sciences centres on studying the nature of networks of international organizations of all kinds, whether they be legal, cultural, and diplomatic, or scientific, commercial, and sporting. Relatively small nations and organizations can have profound influence on world affairs. Sometimes this can be tracked to their strategic or cultural importance or to dominant individuals. However, substantial and sometimes less visible influence often stems from the way they are placed within relevant networks and how they draw from and feed into these webs.

It is fair to say that the first genuine *problem* in networks dates to the eighteenth century when the famous Swiss mathematician, Leonhard Euler, showed how to solve the now celebrated riddle of the Bridges of Königsberg by finding a simple general principle that dealt with any question of that kind. But more of that later. This does alert us however to the fact that networks have been studied from the mathematical viewpoint for centuries. None the less, it is striking how their relative importance keeps growing and growing. In part this is due to examples of networks springing up in modern life—we need look no further than the internet to find a massive and important instance of a network that has come into being almost spontaneously. This network pervades most aspects of the modern world and has taken on a life of its own. The internet acts as a vehicle for another network, the World Wide Web. These networks differ in two ways, one physical and the other mathematical. The Web is visible but intangible and floats on top of the internet, which is a physical array consisting of routers and their connections. The Web is also a *directed* network for there are links directed from one page to another, but not necessarily in the reverse direction. This gives it a very different character from networks in which all connections are mutual and two-way.

It all goes much deeper than that however. Professional mathematicians have tended to have a similar reaction to that of the general public to the underlying idea. The notion of a network of connections is so simple and natural that there looks to be not much to it. To be sure, even in the eighteenth century Euler showed that even a simple example can yield an interesting problem. All the same, it was felt that the depth and interest of the mathematics involved could hardly be on a level comparable with really serious science, such as that which explains how the Earth and the Heavens move. Since the time of Isaac Newton, calculus, the mathematics of change and movement, has been a well-spring of scientific inspiration and was seen as the heir to classical Greek geometry, representing the pinnacle of mathematical practice and sophistication. Indeed Leonhard Euler himself perhaps did more than anyone who has ever lived to develop the methods of Newton, the so-called differential and integral calculus. By comparison, problems about networks were regarded as a poor relation, little more than recreational puzzles, fit only for those who could not contribute to the really tough stuff.

Networks, however, spring many surprises. And they truly are surprises because no one would expect objects with virtually no mathematical structure to yield anything of interest. After all, a network is *any* array of points on a page with lines drawn between some of them in any fashion at all. The idea would seem to be far too general to yield anything that went much beyond the obvious. However, there is a whole world to be explored by those prepared to search and the results have consequences for real networks of people and telephone lines. For instance, at any party that ever there was, or ever will be, or ever could be, there will be two people with the same number of friends at the gathering—this, and many results like this, are unavoidable consequences of the nature of networks, as we shall soon witness.

Part of the trouble has been that mathematics itself has been slow to wake up to what was happening. Problems about networks keep arising irresistibly, even when you are not looking for them. I myself spend a lot of time on my own speciality that is a certain area of algebra. What has happened in my own field has been mirrored elsewhere. Certain intractable problems have arisen and, in the end, progress is only made when they are represented in terms of networks whereupon it transpires that what is holding you up is a question about whether or not certain patterns can or cannot arise in a network. No use sneering—it turns out that nets were really what you have been studying all along.

If your own research topic, stripped of its pretensions, can be cast in terms of networks, you can feel taken down a peg if you had believed that your work was far above such mundane matters. After all, anyone can understand a network. Indeed that is one of the attractions of the subject. It is immediately accessible to everyone and having an encyclopaedic mathematical knowledge often does not help that much in the real world problems that arise. Sometimes though, the reverse is true. Some problems in the theory of networks have been tamed by the use of very sophisticated mathematics and you will see glimpses of why that should be so as our story unfolds. Nonetheless it has not been possible to subsume the theory of networks into an environment where all the problems that arise can be, in principle, dealt with by a standard body of mathematics.

It is true that we may generalize the idea of a network in several ways. There is a branch of mathematics known in the trade as *matroid theory* which, for example, includes networks under its umbrella. When problems of matroids are solved, they automatically have consequences in the theory of networks. This does not mean, however, that the study of networks has been genuinely superseded. To explain with a more common analogy, we all know that any ordinary counting number, such as 6, can be represented by a fraction, 6/1. We have a very good understanding of fractions, but that does not mean that the theory allows us to solve all important problems about whole numbers. Placing a question in a wider setting does not automatically augment your comprehension of it. Indeed it might be a misguided distraction that will not help at all.

In this opening chapter, we begin with the simplest kind of network, that which resembles a tree. At first glance, some of our motivating questions may seem to have nothing to do with networks of any kind but, as you will see, the connections lie there, not far below the surface.

Trees

Before we begin explaining things using trees it is best to pause to say just what we mean by this term, *tree*. The name suggests that the picture of a tree should have a trunk with branches and twigs stemming from it. That is largely the case although there is not always an obvious trunk to a mathematical tree. One mathematically precise way of defining a *tree* is a network where there is exactly one path between every pair of nodes. Equivalently, a tree is a connected network (one that comes in one piece) that is free of cycles. Yet another way of looking at trees is as the networks that are connected and have one more node than edges.*[1]

Figure 1.1 is the family of all trees with six nodes (and so five edges). You can check for yourself that there is only one possible tree with each of one, two, or three nodes. However, with four nodes you will be able to draw two different trees, with five nodes there are three trees and, as you can see, with six nodes there are six. After that it gets complicated: with seven nodes there are also six distinctly different trees but with eight nodes you can find twenty-three in all. Going beyond this, the numbers grow quickly although rather erratically. There are 104,636,890 different trees that can be drawn with just twenty-five nodes.

[1] We will not stop to demonstrate mathematical facts such as these. However, an asterisk indicates that the matter is dealt with more fully in the final chapter, 'For Connoisseurs'.

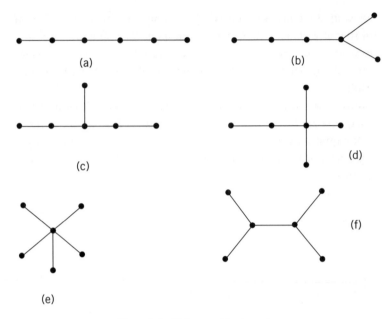

Figure 1.1 All trees with six nodes

Nevertheless, by hunting through systematically, you can find them all. For example, for six nodes, there is just one tree with a longest path of length 5 (a), two with a longest path of length 4, (b) and (c), two where the longest path has length 3, (d) and (f), and just one that contains no paths longer than 2, that being (e). We see there is an essential difference between (b) and (c)—both have a node of *degree* 3, meaning there is one node that has three neighbours, but in (b) this node is adjacent to two endpoints, which is not the case for tree (c). By the same token, (d) has a node of degree 4 whereas (f) does not, so these pictures represent different trees. Indeed these observations suggest another way of categorizing the trees that we meet: tree (a) is the only one with no node of degree more than 2; trees (b), (c), and (f) have a maximum degree of 3, while (d) is the only one where the maximum degree is 4. If we introduce a node of degree 5, then all we can draw is the 'star' we see in network (e).

You might convince yourself for a time that you have drawn another tree on six nodes that is different from the six listed above. However, you will find that if you imagine your tree made of five stiff rods linked together by movable joints, you will be able to manipulate your tree to fit one of the above pictures. The simple classification type arguments of the above paragraph should help you quickly to identify the right candidate.

One aspect that comes to the fore when discussing networks is that many simple ideas arise that have quite natural names. Up till this point we have used the words node, edge, path, cycle, connected, and degree, rather casually. Mathematicians fuss a little about the precise meaning of these words (although they don't always quite concur as to which word means what). For example a *path* is not allowed to go back on itself—that is to say, in a path we are not allowed to traverse an edge and retrace it in the opposite direction later; if you do then your 'path' becomes a mere 'walk'. We also distinguish between a *circuit* and a *cycle*: a circuit returns to its starting point but is permitted to cross itself along the way, as in a figure-of-eight. In other words, in a circuit, nodes may be revisited although edges are used only once. A *cycle* on the other hand, also known as a *simple circuit*, does not cross itself as you traverse it.

A serious study of trees would require us to explain precisely what we mean when we say that two pictures represent the 'same' tree. However, such precision, although necessary in the long run, can wait for the time being as we are keen to visit some real problems.

Before moving on to examples, however, it is worth taking a moment in order to illustrate the terms introduced. The network of Figure 1.2 serves this purpose (without having any particular meaning of its own). The network has three *components*, by which we mean connected pieces, although the middle component merely consists of an *isolated* node d with no edges connected to it. Between nodes a and b there are *multiple edges*, as there are between f and h. We say that two nodes are *adjacent* if they have an edge running between them and two edges are *adjacent* if they share a common node. We also say that an edge is *incident with* the nodes at either end. The *degree* of a node is the number of edges incident with it so

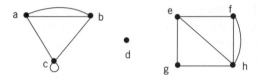

Figure 1.2 A typical network

that the degree of node a is 3, that of h is 4, while that of d is 0. At c we see a *loop,* which is an edge that starts and finishes at the same node so that the degree of c is also taken to be 4 as the loop at this vertex counts twice. We call a node *even* if it is of even degree (and 0 is included among the even numbers); otherwise the node is *odd*.

An example of a path in this network is $e \to g \to h \to f$, although there are two edges possible for the passage from h to f. For a walk to be called a path we normally insist that there are no repeated vertices, that is to say a path does not cross itself. A *trail* on the other hand is allowed to do this and an example of a trail that is not a path is $e \to h \to f \to e \to g \to h$. A closed trail is a *circuit*: for example $a \to b \to c \to c \to a$ is a circuit as it returns to its starting point (from where ever you begin).

The most general type of passage through a network is called a *walk* for here we are allowed to repeat vertices and edges if we choose. An example of a walk that is not a trail (and so not a path either) is $e \to g \to h \to g \to e \to f$. An example of a cycle is $e \to g \to h \to e$. As you see, there are many terms involved in traversing a network and their usage is not completely standard. Just occasionally, then, it becomes necessary to pause and spell out exactly what you mean when talking about traversing a network in a particular way, making it clear what you are and are not allowed to do.

A cycle is sometimes called a simple circuit, while a *simple network* is one in which no loops or multiple edges are permitted. In many books, especially mathematics texts, the term *graph* is used in place of network. Indeed the entire subject is referred to in the official literature as *graph theory*. This takes a bit of getting used to as the word 'graph' is also taken to mean the plot of a function, not just in mathematics but in everyday language. The term network is often reserved for graphs that carry additional information on their vertices and edges, especially when the network is a model of a real set of connections of real objects so that edges might represent pipelines or economic costs. We shall stick to the term network or occasionally just *nets*.

Chemical isomers

Despite their simplicity, no one seemed to have recognized trees as objects worthy of investigation until the mid-nineteenth century. Often it takes a first-rate scientist to appreciate that a topic that looks so simple as to be beneath serious notice is truly important. The first person to study and use trees seems to have been Gustav Kirchhoff in 1847 when devising the laws that govern electrical circuitry. We will return to this direction in Chapter 7. On the other hand, the famous British algebraist Arthur Cayley began studying trees around 1857. He was motivated not only by curiousity but rather was prompted by particular problems. He was first led in this direction by something rather technical, although mathematically important: the nature of the Chain Rule of differential calculus when applied to functions of several variables.

The same diagrams that we now call trees arose in a totally different setting, which is that of enumerating the so-called *isomers* of the saturated hydrocarbons—that is molecules with chemical formulas of the form C_nH_{2n+2}, where C and H stand respectively for a carbon and hydrogen atom. Essentially Arthur Cayley (1821–95) was trying to find all the trees in which every node is either an endpoint or has degree four, as four is the *valency* of carbon—the maximum, and chemically preferred, number of bonds a carbon atom may make with other atoms. For this historical reason, the degree of a node in a network is still referred to as its valency in some books. The five smallest saturated hydrocarbons have their trees displayed in Figure 1.3: the endpoints are occupied by hydrogen atoms while the other nodes are taken up by carbon.

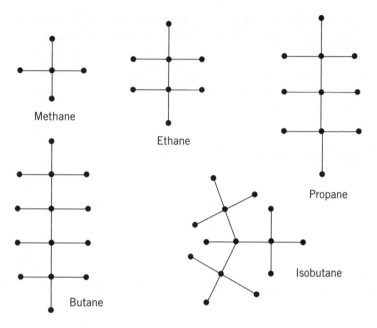

Figure 1.3 Trees of the simple saturated hydrocarbons

Lying liars and the lies they tell

A situation in which tree diagrams naturally arise is in the analysis of a procedure where a succession of decisions is made leading to the process evolving along a path with many forks. This is just what happens in logic puzzles and one of the most elementary of these puzzles concerns two tribes, one truthful and the other liars. The standard version is that you meet a native at a fork in the road and you need to find which way to go. You cannot tell by appearances if he is truthful or a liar and the rule is that you may ask only one question. There is obviously no value in asking which road you should take and the standard solution is to ask, 'Which road would a member of the *other tribe* direct me along?' The native, whether a liar or not, will then indicate the *wrong* road and you then reply by saying, 'Well, we know not to pay any attention to them' and set off down the other.

I have always thought that this course of action was a little unwise as it carries with it the double risk of both confusing yourself and offending the native. You can, after all, get the truth straight away by simply saying to him, 'Which road would you tell me to take if I asked you?' You will then be shown the *correct* path as a lying native will be forced to lie about his lying. This question guarantees you a true answer while providing the diplomatic bonus of not having to bring up the potentially unwelcome subject of the other tribe.

Indeed this trick applies to any query you care to put to a native of this strange land. As long as you phrase your questions in the style, 'What would you say if I asked you...' you will never be misled. However, if the natives start speaking of their own accord, they are liable to cause mischief.

Suppose you meet up with three of them, *A*, *B*, and *C* who are not too forthcoming about their tribal allegiances. Indeed *C* refuses to speak at all, *A* merely offers the coy remark, 'Some of us are liars,' while *B* volunteers, '*A* would call *C* my tribal brother!' What are we to make of this trio? From which tribes do *A*, *B*, and *C* come?

How can we analyse a situation like this? We can display the possibilities on offer as a tree. Each of the three natives has two possibilities for his tribe so the full set of tribal allegiances of the trio consists of $2 \times 2 \times 2 = 8$ possibilities. We exhibit all eight in the network of Figure 1.4.

In deference to the traditional setting for this problem type, I will continue to refer to the two types of people as 'tribes'. In some modern treatments, this place of logic and confusion is referred to as 'The Land of Knaves and Knights', a creation of the American logician Raymond Smullyan, where the Knaves are the liars and the Knights are always truthful. Of course, it all amounts to the same thing.

The node at the top is labelled *A*, those on the second level *B*, and the four at the third level, *C*. Each line, or edge as the lines of a network are more often called, is labelled by either *T* or *L*: those to the left labelled by *T* indicating that the native represented by the corresponding node above is truthful while those to the right indicating the opposite, that being they are liars. The procedure we adopt is that of a *tree search*: each path from the top node (the *root*) to

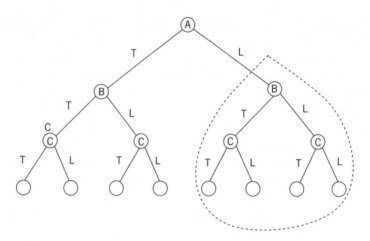

Figure 1.4 Tree of all possible tribal allegiances

an endpoint represents a potential resolution of the tribal allegiance question. We check all eight paths from the root to the bottom level and discard any that contradict the facts as given in the problem. If we are lucky and there is but one *successful path*, that will correspond to a unique resolution of the tribal membership problem. Conceivably, however, there could be several successful paths, in other words several solutions consistent with what the natives have told you. This would mean that you have not yet learnt enough to be sure of their respective tribal memberships.

With a little good fortune, you will not have to conduct a full search of all paths. A path may lead to contradiction at the first or second stage, allowing us to eliminate more than one path in a single stroke: if an edge gives contradiction we can 'prune' all of the tree from the corresponding node downwards. Indeed that is what happens here when we consider the statement of A. He tells us that some of the trio (that is, at least one of them) are liars. This must be true, for if it were false that would make A a liar, showing that A in fact spoke the truth! The only way to avoid contradiction then is for A to be truthful and so his statement is also. It follows at once that the right-hand four paths in our tree of possibilities, encircled by a dotted curve in the diagram, are eliminated as they each begin with an edge labelled L emanating from A, which represents an impossible situation. Only the possibilities represented by the left-hand side of the tree are still standing.

Let us now looks at the other four paths in turn. There are two paths beginning with TT. In this case B is truthful as well, and so his statement that A would declare that C and he are brothers would also be true. However, the truthful A would only say this if C and B were both truthful, eliminating the TTL path, leaving TTT. However, if all three were from the truthful tribe, the truthful A would not have said some of them are liars, so this branch also yields contradiction. Therefore both TTT and TTL are pruned from the tree of possibilities.

The only remaining paths are TLT and TLL—in other words we are sure that A is truthful but B is a liar and so only the tribal allegiance of the silent member of the trio is any longer in doubt. Now since B is a liar, his statement concerning A must be false, as this applies to everything he ever says. Suppose now that C were also a liar. Since B's claim about A is false, A would *not* say that B and C were of the same tribe—but since this pathway represents a scenario where B and C are of the one tribe (lying brothers), this would mean that A would also be lying, something he just never does. Hence we have reached another contradiction and so C is a truth teller and the remaining path, TLT, is the correct one.

We should check that TLT is really consistent with what we have been told, just in case we have somehow misheard what A and B have said. In this pathway, B is lying about what A would say: A would *not* say that B and the taciturn C are brothers, and since B is a liar and C is not, this would be consistent.

In summary, we conclude that B is the only liar in the group and that C, should he ever choose to speak, will speak the truth.

Of course, in any legitimate scenario like this one there has to be at least one solution to the tribal membership question, as otherwise the scene could never have come to pass in the first place. If, however, one just imagines a collection of natives saying this and that, there could be unavoidable contradiction in the problem that may or may not be obvious on a first hearing. For example, you would never meet a native of this strange land who would say candidly, as you or I might, 'Yes, I do lie sometimes.' Even the most modest member of the Truth Tribe could never say this, as it would be a lie, while a member of the lying tribe could never say it either as it would be the truth!

Problems like the previous one on our two tribes can be quite fun but it is crucial to get the meaning of the wording absolutely clear. Under one interpretation, the *TLL* solution is also feasible. It may be that A would not say that B and C are brothers simply because he may refuse to make any such statement even though it is true. If that were so, even though B and C are brothers, B is (technically) lying when he says that A would call them brothers. Being in the company of liars, A may find it circumspect to say nothing more than he already has, while B might be a more cunning liar than we gave him credit for.

In setting up our tree of possibilities, I began with A at the top because I analysed the situation first by looking at what A was saying. However we could similarly analyse all this beginning with what B has to say, an exercise you might care to try yourself.

This is all reminiscent of the ancient paradox of Epimenides of Knossus (*c.*600 BC) who famously asserted that 'Cretans always lie,' even though he himself was a Cretan. If Epimenides had only said that Cretans are liars, simply meaning that they could not be trusted to tell the truth, then the statement could well have been true and apply to himself also. However, he made a stronger claim, that being that Cretans all belong to a Lying Tribe, and this is paradoxical as it should then not be possible for a Cretan to implicitly call himself a liar in this way. The statement of Epimenides therefore leads to contradiction if we assume that it is true but does it also yield contradiction if we take it to be false? It would seem that the ordinary commonsense meaning can be reconciled with logic by assuming that Epimenides is a liar, at least in this instance, but that some Cretans are not.

However, this leads to a strange conclusion in itself, for we then seem to be claiming that we now know, beyond any doubt, a certain fact about Cretans from a single statement made by one Cretan that is in any case false. This surely does not make sense—we have never come across any of the other Cretans from 2,600 years ago and we know nothing about Epimenides but this one inconsistent statement, so how can we know anything about them with any certainty?

There seems to be more to be said about this, however we look at it. The twin spectres of self-reference and unknowable existence will raise themselves again later when we consider a famous theorem of Brouwer on fixed points, which in turn returns us to the question: Is it reasonable to take for granted that every statement we could ever make is either true or false?

We close this chapter with a couple of further examples involving fibbing. Can you dispense justice in the following situation? Four children are playing when the window is broken. Alex says that Barbara did it, Barbara says that Caroline did it, while both Caroline and David say they didn't see what happened. Assuming only the guilty child is not telling the truth, who broke the window?

Returning to our natives: suppose that you meet up with a pair of them and the first says that the other is a liar to which the second responds by claiming that they are from different tribes. What tribes are they from?

As mentioned before, sometimes you may only be able to infer a partial solution to the problem from what the natives tell you. A scenario that arises all too often in Courts of Law is where two defendants accuse one another of the crime. All that you can be sure of is that at least one is a liar but the jury cannot convict either of them from this evidence alone. If you meet two tribesmen who assert that the other is a liar you can be sure of a little more. Since the tribesmen *always* lie or *never* lie, you can deduce that you have exactly one member from each of the two tribes, but symmetry precludes you from deciding which is which!

As a final mindbender from the Land of the Two Tribes, suppose you come across a whole string of them—*A* says *B* is a liar, *B* says that *C* is a liar, *C* says the next in line is a liar, and so on, right down to the last man who says that they are all liars except for him. What is going on here?*

2

Trees and Games of Logic

M any games and puzzles, perhaps more than you realize, are tree searches. Examples include games like *Mastermind* and puzzles such as the modern creation known by the Japanese name of *Su Doku*,[1] although this puzzle type seems first to have appeared in the US-based *Dell* magazine many years earlier. The most difficult are the board games such as checkers and chess.

Familiar logic games

The simplest is tic-tac-toe or noughts and crosses. This pastime is complicated enough to be a real game yet simple enough for we humans to master it completely.

The first player marks his cross in any of the nine squares while the second player counters with her noughts. They continue until one or other gets three of their symbols in a line (horizontal, vertical, or diagonal) or until all nine squares are occupied without this happening in which case the game is drawn. It is obviously a big advantage to go first in this game, yet, if the second player is careful, she can avoid defeat and even win if the first is careless.

Many other popular games are the same as tic-tac-toe in that players move in turns and each battles to take the game down a favourable path in the tree or at worst avoid an unfavourable path.

---

[1] Or simply *Sudoku*, the literal translation of which is 'single number'.

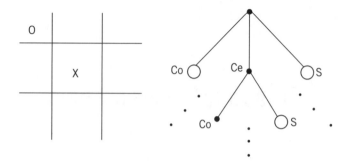

Figure 2.1 First pair of moves for noughts and crosses

Generally, though, the trees are so big that it is not possible for a human and sometimes even a computer to be absolutely safe by checking all final outcomes at every stage.

The full tree for tic-tac-toe is still quite complicated. It consists of an inital node with nine levels below, the first corresponding to all possible game positions after Cross has his first move, the second to all positions that could arise after each player has had one move, and so on. Of course some paths down the tree finish early because they lead to a game in which one or other of the players has won.

Although there are nine initial moves for Cross, there are really only three genuinely different positions the game can adopt after the first move: either the first X has been placed in a corner, in the centre, or in a side square; for that reason the initial node has but three offspring nodes, marked in Figure 2.1 as Co (Corner), Ce (Centre), and S (Side) respectively. If Cross does choose the centre square first, there are only two truly different responses by Noughts: either she goes for the Corner or a Side square, and both possibilities are drawn in the partial tree in the diagram. The shaded nodes correspond to the game that begins as in Figure 2.1, taken to the second stage as the picture shows. As we know from experience, Noughts has enough control over the game so that, on her turn, she can always direct the game down a path in the tree that can avoid a node in which Cross wins, although a thorough analysis of all possibilities is required to verify this. All the same, it is not too hard to do.

One game that is perhaps a bit too tough for most humans but which computers can totally master is *Connect Four*. In this game opponents take it in turn to drop red and gold coloured plastic discs into one of seven possible slots of a vertically mounted board. Each slot can support a column of up to six discs. On each turn, a player has at most seven choices of column into which to drop his disc. Like tic-tac-toe, you win by being the first to create a row, this time of length four, in your colour. By symmetry of the board, the initial node of the game tree has just four offspring, but after that the number of possibilities mushrooms giving a game with a huge number of variations. What is more, this tree goes much deeper: there are not merely nine levels below the starting node but 42, corresponding to the 7 × 6 places on the board, each of which can represent the placement of a disc, and so a turn of one player or the other. Once more, the opening colour (red) has an advantage but, seeing as *Connect Four* is a relatively long game, we might expect that the advantage is not decisive. Surprisingly, however, it turned out that *Connect Four* is a forced win for red: with best play, the player who goes first can always ensure that he wins—there are now computer programs that win every time in *Connect Four* if they are allowed to go first.[2]

The Chess Tree, like that of the two previous games, consists of one node for every possible position that could arise in the course of a game of chess. Each node has a number of edges emanating from it, one for each possible move from that position. The number of edges coming from a node varies between 0 when a game has ended or reached a stalemate where no move is possible,[3] up to about 30— if you count the number of moves available to you at any point in a chess game, it is rare to find you have more options than about this number. The playing of any one game of chess involves following a path down the tree from the common starting position of all games to some endpoint of the tree representing a completed game. Your

---

[2] This was verified independently by two men: the Dutch artificial intelligence researcher Victor Allis and the Californian computer scientist James D. Allen.

[3] According to the rules of chess, if a player, on his turn to move, is not in check but has no legal move, the game is at a *stalemate*, and is declared a draw, an outcome that arises in real games on odd occasions.

task at each stage is to pick a good move—one that leaves you in a section of the tree where you can win or at least draw. However, the choice of move at the odd numbered levels, the first, third, fifth, and so on are the preserve of white (who always has the privilege of the opening move) and the even numbered levels belong to black. Ideally you want to lead the game down a path in which you can ensure that you win. If this is achieved players say that, from this particular position, you have a *forced win*. If you reach a node where you have a forced win, it does not mean that all endpoints from this point on lead to you winning, for it is virtually always possible to squander a winning position. It does mean, however, that whatever branch your opponent chooses from this point onwards, you can, by making a suitable move, direct the game down a branch that finishes with his king being the victim of checkmate. Presumably this does not apply to the initial position, but that is something that has not been proved.

Checkers is a similar game to chess, indeed it is played on the same board. The tree representing all possible checkers games also extends to many levels although the branching at each node is less vigorous. Compared with chess, at a typical position in the game, the player has fewer alternative moves from which to choose. In both chess and checkers it is also true that most moves on offer are obviously bad and this will be clear to any experienced player. For that reason it is possible for the best checkers players to look many moves ahead, they say up to thirty at times. In chess, even the grandmasters are not mentally searching down the tree of all possibilities to that kind of depth: the top players generally only 'see' the board about half a dozen moves ahead. A game of chess is sometimes likened to looking across a broad field while in checkers the players are peering into a deep well.[4]

Another member of the club of ancient board games is the oriental pastime of *Go*, a game of competing black and white counters played on a large 19 × 19 board. Each player jostles for space as they connect with their own counters to surround and remove those of

---

[4] A group based at the University of Alberta now claim to have solved checkers in that their program is unbeatable. In particular, this shows that the initial position is a drawn one. See www.cs.alberta.ca/~chinook/news/.

their opponents. The tree of possibilities here must be enormous and as yet it is claimed that no computer program can compete with the true masters. Nonetheless, it is a game that looks ripe for attack by a powerful computer primed with a cunning program. This seems even more plausible as computers now reign supreme in the similar-looking game known as *Reversi* or *Othello*. Othello is named after Shakespeare's famous Moorish character who treacherously had the tables turned on him by the two-faced Iago. This game is played with black and white counters on an 8 × 8 board where besieged pieces are inverted and thereby have their colour reversed to that of their opponent.

The trouble with trees like the chess tree is that they are so enormous it is impossible to conduct a full search. Even modern computers have no hope of searching anything like the whole chess tree. Both human and computer chess players rely on *heuristics*, that is rules of thumb, to guide their local search of the tree: a game position represents a particular node in the chess tree. Whether you are a human or a machine you will be able to examine completely only a portion of the tree below this node, which corresponds to all possible 'near futures' in the game. Ordinary players rarely look more than a couple of moves ahead. Grand masters sometimes directly take account of all possibilities up to the next dozen moves, although brute force calculation is only one weapon in their arsenal. Computer programs on the other hand do try to handle as much of the tree as they can cope with. However, to be effective, they also have to have simple rules that tell them when not to bother searching one branch of the tree while persevering with another. One rule for instance might be, if a series of future moves ends with a piece being removed from the board, do not halt the search of that branch but look down one more level to all configurations. Eventually, however, the computer will have to give up the search and choose one move, that is one branch of the tree, which is rated as the best by some measure that the programmer has built in.

Computers and humans are both very good at chess but their approaches are different. The more successful programs have not tried to create machines that mimic human thinking but rather exploit the strength of the computer, which is massive memory recall and direct computational power. It is an interesting facet of modern chess to have two fundamentally different kinds of player with contrasting strengths and weaknesses. A human player might seek to 'rattle' his computer opponent by negating its computational skill. This could be done by the human using openings of which they have special knowledge or making the game very messy and complicated and so give the machine a scrappy fight. If the human can put the program in positions that its decision rules do not cope with well, the computer is liable to do something dumb. To the amused onlookers it will appear that the poor machine has got flustered. The computer on the other hand should seek to control the number of possibilities so that its computational power would allow it to gaze much deeper into the game than its human opponent who would then be left floundering and out of his depth.

A classic example where a suitable tree search quickly solves a conundrum is that of the Counterfeit Coin. There are nine coins, one of which is fake and can be detected because it is a bit too light compared with the genuine article. You have a set of balances you can use to identify the dud coin but the task is made more of a challenge by you being confined to just *two* weighings before you flush out the imposter. A tree with just three levels does the job (Figure 2.2).

The idea is to use the scales to eliminate two thirds of the possibilities at each weighing. We imagine the coins numbered 1 through to 9 and we compare the collective weights of the sets {1, 2, 3}

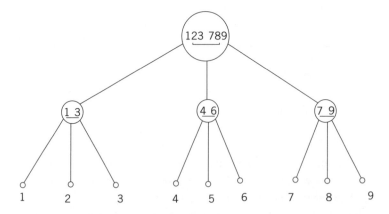

Figure 2.2 A tree to detect the counterfeit coin

and {7, 8, 9} as represented by the first node of the tree. At each stage we move to the left down the tree if the left-hand side of the scale is the lighter, to the right in the opposite case and, if the scale remains in balance, we move down the middle. After two weighings this will lead you to the false coin. For example, if 8 is the fake, the first weighing will see a light right-hand scale, so we know the culprit is one of 7, 8, or 9. We move down the right-hand branch of the tree and compare the weights of 7 and 9. They will of course be in balance and so we move down the central branch emanating from that node and conclude that it is coin 8 that is the fake.

This procedure can be applied to any number of coins with a similar result based on a *ternary* tree, one in which each node has three branches.* The problem is much tougher if it is not known whether the bad coin is lighter or heavier, just that its weight *differs* from that of a true coin.[5]

Exotic squares and Sudoku

The most recent logic puzzle to catch the global imagination is the Japanese game challenge of *Sudoku*.[6] We introduce these puzzles by way of a short digression. The game of Sudoku is based on the idea of a *Latin Square*, which should not in any way be confused with *Magic Squares*.

A magic square is a square array of numbers in which every row, column, and diagonal sum to the same number. Constructing a magic square is a problem in arithmetic rather than straight logical analysis.

Figure 2.3 shows two of the more famous magic squares. The first is known as the Lo-shu and was discovered in China some thousands

---

[5] To see these trickier versions tackled you can browse the web page <http://www.iwnteiam.nl/Ha12coins.html>.

[6] The history of Sudoku is now well documented: its modern origin was in the States although it seems to have enjoyed a brief popular life in France around the end of the nineteenth century. This is not unlike the history of Reversi, which originated in nineteenth-century England, but only became widely popular after being taken up in Japan in the 1970s.

16	3	2	13
5	10	11	8
9	6	7	12
4	15	14	1

4	9	2
3	5	7
8	1	6

Figure 2.3 Two famous magic squares

of years ago. The numbers 1 through to 9 are arranged so that every line and diagonal sum to 15. This is really the only 3×3 magic square featuring the first nine numbers: any other you might come up with can be realized by taking the Lo-shu and rotating it about its centre or reflecting it about its diagonals or sides.

When we pass to larger squares things get more complicated. The 4×4 square appears in a picture engraved by Albrecht Durer. The magic number that represents the sum of every line in this case is 34. (If we use the numbers 1 through to 16, the common line sum has to be 34, as that is one quarter of 136, the sum of all the numbers in the square.) The date of the engraving is also there for all to see, 1514, but the square has other magical features as well. Each quadrant also sums to 34 (for example $7 + 12 + 1 + 14$), as do the set of four numbers that make up the middle of the square. Surprisingly, there are more symmetries still: if we glue the top edge to the bottom, the 4×4 square we see also respects the magic sum: $3 + 2 + 14 + 15 = 34$, and the same happens when we stick the vertical sides together: $5 + 9 + 12 + 8 = 34$, and the four corners add up to 34 as well.

A Latin Square on the other hand displays a symmetry that is not about arithmetic but is more about balance: a Latin Square is an $n \times n$ array where each of n distinct symbols appears exactly once in every row and every column. The symbols used are often the numbers 1 through to n but that is just for convenience as they

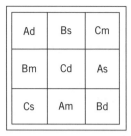

Ad	Bs	Cm
Bm	Cd	As
Cs	Am	Bd

Figure 2.4 Family Graeco-Latin Square

can be any symbols you fancy such as letters of the alphabet, signs of the zodiac, or even colours.

It is not hard to make a Latin Square of any size n: just write down the numbers 1 through to n for your first row, do the same for your second row except this time start from 2, start the third row from 3, and so on, writing each row in cyclic order. And there are lots of other ways of dreaming up Latin Squares.*

A more interesting challenge is that of finding so called *Graeco-Latin Squares*, which we introduce by way of a problem. Suppose we have three families, named **Adams**, **Baker**, and **Collins**, which each consist of a **Mum**, **Dad**, and **Son**.

Can we arrange all nine in a square so that each family and each type of family member is represented in every row and every column?

This seems to be quite a tall order but it can be done as is seen in Figure 2.4. Each person carries two labels: their family name and the type of family member they are and each of these labels is present in every line. (But not every diagonal—the Collins family have a diagonal all their own while the Dads monopolize the opposite diagonal.)

Can we extend this to larger squares? In the 4 × 4 case we might imagine four families each with two parents, a son, and a daughter. Before we answer this it might be best to pause to examine what it is we are attempting, which is to place two Latin Squares, one on top of the other, in a certain way.

Imagine we had for instance a 4 × 4 Latin Square made up of the first four letters of our Roman alphabet and another made from the first four letters of the Greek alphabet, α, β, γ, δ. If we were to superimpose the Roman Square on top of the Greek Square each of the letters A, B, C, D and α, β, γ, δ would appear just the once in each row and column of the square. The Graeco-Latin challenge now arises as follows. For any such $n \times n$ square the number of possible pairs of one Roman letter with one Greek letter is also $n \times n$: in the case of a 4×4 square there are 16 pairs possible featuring one Roman and one of the Greek letters.

Is it possible to choose the Roman and Greek squares so that every pairing of letters occurs in the superimposed squares?

We have already done this for the 3×3 square. It turns out that the answer is 'yes' for squares of orders 3,4, and 5, where by *order*, I mean the number n for an $n \times n$ square. Figure 2.5 exhibits solutions for the $n = 4$ and $n = 5$ cases, although instead of Greek letters we have used lower case Roman letters as they serve just as well: in each row and column each capital and each lower case letter appears exactly the once and every possible pairing from the two sets occurs somewhere in the array.

Beautifully balanced arrays like this are very useful in the design of real experimental trials for they allow, for example, the thorough mixing of pairs of treatment types over a field.

In the eighteenth century, Leonhard Euler, who we will meet on more than one occasion in this book, showed how to make

Aa	Bc	Cd	Db
Bb	Ad	Dc	Ca
Cc	Da	Ab	Bd
Dd	Cb	Ba	Ac

Aa	Bd	Cb	De	Ec
Bb	Ce	Dc	Ea	Ad
Cc	Da	Ed	Ab	Be
Dd	Eb	Ae	Bc	Ca
Ee	Ac	Ba	Cd	Db

Figure 2.5 Graeco-Latin squares of orders four and five

Graeco-Latin squares of order n for any odd number or any number that was a multiple of four. However the other even numbers, 2, 6, 10, 14, ... would not cooperate and the corresponding Graeco-Latin squares remained elusive.

It is obvious, and you will soon convince yourself should you try, that there is no 2×2 Graeco-Latin square. Euler then set the challenge for $n = 6$ by way of his *Ranks and Regiments* problem. There are six ranks and six regiments and they want to send a marching square of 6×6 through the town with every rank and every regiment represented in each row and file and with every regiment being represented by each rank (and so, in consequence, vice versa). We see that Euler was in effect asking for the construction of a Graeco-Latin square of order six. He conjectured that it simply could not be done and he went further to suggest that there were no Graeco-Latin squares of order n if the number n has the form $4m + 2$.[7]

There the matter stood for over a hundred years until in 1901 Gaston Terry showed that Euler was right about his ranks and regiments and there simply was no Graeco-Latin square of order 6. This strengthened the general belief that Euler had been right all along but no further progress was made for over fifty years. Then, in 1959, some counterexamples to Euler's Conjecture were discovered and soon an example was found of a Graeco-Latin Square of order 10—this proved to be quite a sensation at the time, featuring prominently in the *New York Times*. It did have the making of a good story as the papers could not only explain the question but could print the solution for everyone to see. This psychological breakthrough led to an avalanche and in 1959 Bose, Shrikhande, and Parker showed that Euler was, on this occasion, almost as wrong as can be. There *are* Graeco-Latin squares of any order greater than 2 *except* for $n = 6$. In the entire infinity of numbers, only Euler's Ranks and Regiments are fated never to find a solution.

Returning to the new fashion of Sudoku problems, they are all based on 9×9 Latin Squares, and so the symbols used are

[7] These numbers are known troublemakers—they are the only ones that are *not* the difference of two squares.*

conveniently taken to be the numerals 1 through to 9. The puzzle-setter gives the player a partial Latin Square to complete with the added proviso that *each symbol has to appear in each of the nine* 3 × 3 *squares* that make up the big square. Like most crosswords, the given display is usually symmetric with respect to the pattern of occupied cells. In the case of Figure 2.6, the picture possesses rotational symmetry through a half turn about the centre, due to its reflectional symmetry through horizontal and vertical axes through the same point. This visually attractive feature was not always present in the original versions known as 'Number Place' in *Dell* magazine in the early 1980s. However, it is not an important aspect of the challenge, as the symmetry does not extend to the structure of the puzzle itself in that the logical interactions between the numbers are not symmetric: for example, in the puzzle below, if you work out the square in the top left corner that does not automatically give you the bottom right-hand corner 'for free' by replicating the corresponding pattern of numbers.

Again, when playing the more difficult Sudoku, the reader is in for a tree search, testing successive guesses until she gets stuck, in which case the player has to re-trace one or more steps up the tree and try another branch instead—it can be a very complicated business. A fine training session can be found on <http://www.sudoku.org.uk/PDF/Solving_Sudoku.pdf>. The setter always guarantees that there is a solution to Sudoku and that it is unique: there is one and only one path down the tree that terminates with a full Latin Square in which every minor box has every digit as well. Old hands of this new sport will not find the puzzle of Figure 2.6 too much of a challenge.*

In general there are lots of ways of completing a Latin square, although it is possible to get stuck. You can reach a stage, as in Figure 2.7, where although you have not broken the rule as yet, you cannot complete the square as there is one cell where, no matter which number you insert, there will be a duplication in its row or its column.

However, this example shows the fastest way possible to foul up the square—there is a remarkable result that if fewer than n symbols

	2		7		4		9	
		5	6		9	2		
1								7
5			4		8			2
		2				6		
8			3		7			4
9								1
		8	1		2	3		
	4		9		5		8	

Figure 2.6 Sudoku Puzzle

are placed in an $n \times n$ square in such a way that no row or column contains a repeat, then it is always possible to complete this partial Latin square to a full one, although perhaps in more than one way. This was first conjectured by Trevor Evans in 1960 but it took twenty years before a proof was found. The proof is only a few pages long but is very delicate and clever—it is considered by some to be one of the prettiest proofs in the world and so merits a place in the compendium of stunning mathematical tricks: *Proofs from the Book*, by Aigner and Ziegler.

1	2	3	4	
				5

⋮

Figure 2.7 A partial Latin square that cannot be completed

When constructing Latin Squares row by row, at any stage we have a *Latin Rectangle*, that is to say an *m*-row and *n*-column rectangle, with *m* no more than *n*, such that no number appears twice in any row or column but, because the square is incomplete, only *m* of the *n* numbers yet appear in each column. The question then arises as to whether it is always possible to keep building the rectangle up to a square or might we become thwarted, with no way to continue. It turns out that it *is* always possible to complete any $m \times n$ Latin rectangle to a full $n \times n$ Latin square and we will be able to explain why in a later chapter. Astonishingly, it is a fact that can be deduced as a consequence of studying the maximum capacity that can flow through a network of pipes.

As a variant on the traditional Sudoku, I can offer you the next puzzle (Figure 2.8) that is a circular version of the same idea. Each of the four rings and **eight** quarter circles are to carry each of the numbers 1 through to 8. The solution is unique and can be found in the final chapter along with some hints on how to solve this problem type.* Although different from orthodox Sudoku, it is similar in that Latin Squares have slipped into the puzzle, even though they are not visible at first glance. Once you appreciate this, you should be able to solve the problem relatively easily, without having to guess and backtrack in the search for the solution.

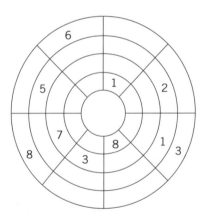

Figure 2.8 Circular Sudoku

It is also a puzzle that lends itself to scaling up or down. Even a three-ring puzzle requires some thought. Five- and six-ring versions are more difficult and, in principle, puzzles with any number of rings can be produced—some more examples and variants are given in the final chapter of the book. For a complete book of these puzzles, featuring half a dozen variations, you can get *The Official Book of Circular Sudoku*, by Peter and Caroline Higgins.

The puzzle requires relatively few given cells in order to determine a unique solution. In this example ten of the thirty-two are given but in some cases as few as nine are required in order to fix the solution. The first Circular Sudoku puzzle was published by the author in the British national newspaper, the *Sunday Telegraph*, on 26 June 2005 and now features regularly in a number of magazines and newspapers. Most of the *Sunday Telegraph* puzzles were based on a five-ring version in which case there are fifty cells and ten symbols. Each ring and each of the ten pairs of adjacent slices has to carry each of the ten numerals 0 through to 9. The least number of occupied cells that can fix the solution is then thirteen, representing 26 per cent of the total number of entries. There is also a handheld electronic version of the game available based on these five-ring puzzles.

One final game to add to our collection is *Mastermind*, as the player trying to crack the code in this game is conducting a true tree search. This differs from the trees describing tic-tac-toe and chess where the levels of the tree belong alternately to one player and then the other as they compete in trying to force the game down a branch favourable to them. In *Mastermind* each node represents a partial state of knowledge for the active player. Each guess takes the player down one more level in the tree to a node which represents an increase in his knowledge (it cannot decrease but he can waste a go and learn nothing). Eventually the secret code will be revealed, although the active player is only allowed ten guesses. In terms of his tree he has to reach a node representing full knowledge of the secret without needing to search beyond ten levels down in the tree, and therein lies his difficulty.

The object of the game is to discover a secret code consisting of four coloured pegs. The colours, which may be repeated, number six types in all: red, white, yellow, green, blue, and black, and are laid

Figure 2.9 Two guesses at Mastermind

down in a particular order. For example, the code may be yellow, black, black, blue. The active player's turn consists of a guess at the code and after each guess the player who chose the code reveals some information—he indicates

(a) how many of the code pegs in the guess are the right colour *and* in the correct position (each is indicated by showing the player a black marker)
(b) how many pegs (occurring elsewhere in the line) are the right colour but are out of position (indicated by showing a white marker).

For example, suppose after two guesses the codebreaker was presented with information as seen in Figure 2.9. From the first guess, we know that we have all the colours right, but all in the wrong places. For our second guess we have rearranged the colours. Since we have left two in the same places, we know in advance that this guess cannot be right but we have extracted more information. Now two are in the right places, and the other two are the wrong way around. Since red and green have not moved, these must be the pair that are wrongly placed, while white and yellow are correctly positioned. The hidden code must then be G W R Y. We got there by going down only three levels in the tree of possible guesses.

In this example we were lucky with the first guess, but a skilled player should be able to crack any code within the given ten moves. For this reason, a harder version of the game is sometimes played where the codemaker may use empty places as well, in effective adding a seventh colour to the game, which then becomes quite tough.

Trees are the diagrams we naturally turn to for a visual display of any hierarchical structure. Invariably it is the relationships represented by the vertical ordering that are important. The left–right ordering of offspring nodes, for instance, often carries little or no information and may be arbitrary.

The trees that arose in the analysis of our games might be more accurately called *root systems* as they grow down rather than up. The term *tree* is universally preferred, however—look for example in any computer manual describing a network of directories and files.

One system of relationships that is displayed as a genuine (upward growing) tree is that of family ancestry, even though as we trace a path up the tree, we are moving back through the past. But that does raise one interesting question:

Why does the size of past generations not keep growing?

Everyone has a biological father and mother and if we were to trace back our direct ancestry we find two parents, four grandparents, eight great grandparents, and so on. This seems totally inevitable but it is just as obvious that it cannot go on forever. The size of each preceding ancestral generation seems to double and it does not take many generations then until we would have an incredible number of ancestors: twenty generations ago we would have more than a million direct ancestors, and, doubling each time, you find that thirty generations back there are in total more than a thousand million people collectively responsible for your birth. Given that a generation is about twenty-five years, we would be led to believe that around the year AD 1000 there were over a billion people on Earth, which most certainly was not the case. On the contrary, we know that there were a lot fewer people in the past than in the present. (A very large portion of all the people who have ever lived are alive today, and the number of people older than you is always diminishing.) So, how can it be that the burgeoning numbers that appear when we look at family trees are somehow avoided?

One thing of which you can be certain is that multiple marriages and marital infidelity are not a necessary part of the explanation. Even if all your ancestors were absolutely scrupulous in this matter and abided by the strictest rules on reproduction—it only being possible inside marriage and no one may marry more than once (widows and widowers remaining so, for example), the same apparent doubling in the generations would still be there.

The explanation is not hard to find but it is not always easy to see as it does not come to light in a typical family tree as far back as most people can trace, which is only a handful of generations. What happens is that, if you extended your complete family tree, eventually some pairs of parents would be listed more than once. This does not happen in most family trees until quite a few generations have been traced.

What must happen, of course, is that as we continue to track back more and more generations, eventually we begin to see the same people popping up more than once; that is to say not every person in the family tree is represented by a unique node, but rather, there is duplication with some individuals turning up repeatedly. How can this come about?

Collapse will begin to occur when we reach generations in which two siblings appear (although not necessarily in the same generation). This pair only give rise to *one* set of parents in previous generations instead of two, and then only four grandparents instead of eight, and so on. After that, duplications will become more and more common until eventually it must be the case that one generation in your family tree has more individuals than the previous one. Although the portion of your ancestral network representing your recent ancestors may be a tree, the entire object is a much more tangled web indeed whose very ancient members are not even humans!

Even in recent geological times our direct ancestors must have been very few in number. We are told that the entire human population almost perished in preceding ice ages and may have dropped as low as a few thousand hardy and heroic individuals. It seems that we are all one big family, a theme that will be taken up next.

3

The Nature of Nets

In this chapter you will be introduced to an array of different kinds of questions that concern general networks that are not just trees but in which cycles and multiple edges are permitted between nodes. At first inspection the problems and queries may not seem to be about networks but the underlying network comes to life as a natural model of the situation in each case. But first we look a little more closely at the reasons why nets are becoming more noticed.

The small world phenomenon

'I've danced with a man, who has danced with a girl, who has danced with the Prince of Wales!' The idea of this was enough to send the girl singing this old song into raptures. However, as has often been observed, this kind of thing is bound to happen once we start to mix with prominent people. Even if we make a point of avoiding celebrities all our lives, it was observed, apparently by the Hungarian writer Frigyes Karinthy as early as 1929, that it is more than likely that any two people on Earth could be linked through a chain of no more than five personal acquaintances. This claim was supported by direct experimentation carried out by the sociologist Stanley Milgram in the 1960s and formed the basis of a popular play, *Six Degrees of Separation* by John Guare in the 1990s. Why should the shortest paths in the global network of acquaintanceship be so very short?

There is a naive mathematical approach that at least explains why the claim is plausible. Suppose that, on average, a typical person has 100 acquaintances. That person's acquaintances will themselves know about 100 people each so that our original person will be acquainted at up to one remove with $100 \times 100 = 10,000$ people. Continuing in this way, we see that if we allow two intermediate people we link to around $10^4 \times 10^2 = 10^6$, which is a million individuals! (The reader will excuse the use of power notation here—10^n just means the number consisting of 1 followed by n zeros.) One step further gives a link to one hundred million (10^8) people and five links allow us to join our original person with 10^{10}, that is ten billion people, more than all the people in the world!

This line of reasoning has already broken down in somewhat the same fashion as our exponentiating family tree did in the previous chapter. The argument assumes that each set of acquaintances that arises is entirely new whereas in practice we would soon be led back to people who have already appeared earlier in the chains of acquaintances. In other words, we are guilty of some double-counting. To be more precise, the rough part of the argument centres on the assumption that every person we meet along the way contributes 100 fresh acquaintances or more. This is the flaw that, in the long run, inevitably undermines any form of pyramid selling or chain letter spread. As in the family tree expansion, however, the truth does not begin to bite hard until quite a few steps of the process have been run through, although long term collapse is absolutely guaranteed. Unfortunately for gullible victims of pyramid selling scams, that can allow more than enough time for the perpetrators of the scheme to stuff their pockets and bolt!

The argument does lend credence however to the expectation that, beginning with one person, the number of people connected to him or her by chains of acquaintance of length no more than five is likely to be very, very large indeed and could include a sizeable portion of the entire global population.

However, the argument given by Karinthy is more convincing and pays due respect to the true nature of the network of human contact. It exploits the fact that a relatively small number of people in the world are especially well connected. His line of reasoning goes via 'hubs'.

Imagine how you might go about finding a short chain of acquaintance between two randomly chosen people, a man in the USA and a young girl in Nigeria, say. A suitable chain could well be made of two shorter chains that link to two very prominent individuals, for example the United States President and the Pope, who we assume have met. The Nigerian girl may well have met a Catholic Bishop who has had an audience with the Holy Father, while our American man's boss might well be an old friend of a prominent business man who has personally met the President. That is all it takes to link our young Nigerian girl to our American gentleman by the short chain we seek.

The evidence seems to suggest that this picture is not too far from the truth. The nature of personal relationships lends to these very short chains. There are features that mitigate against this, however. Friendship links are not random but have a local bias—the majority of people you know are liable to live and work close by, and this factor tends to inhibit the forming of short chains of acquaintance between distant people as will language and other social barriers of various kinds. Indeed a few centuries ago there were large islands of people separated from all the other peoples of the world. Presumably there were no paths of acquaintanceship at all between the peoples of Europe, the Americas, and the South Pacific until the sixteenth century at the earliest. However, in the modern world most of us have a considerable number of friendship ties to individuals who live far away.

This same pattern of short chains arises within particular groups as well. The mathematical community is a prime example as the phenomenon itself is of interest to its members and is amenable to study. The person who was a hub *par excellence* in the twentieth century was undoubtedly the eccentric Hungarian mathematician Paul Erdös who came to the attention of the general public through Paul Hoffman's biography, *The Man Who Loved Only Numbers*. Erdös spent most of his life, certainly the latter half, as a mathematical vagabond, being looked after by his colleagues throughout the world, never spending very long in one place. He simply did mathematics all the time, mainly in the fields of networks, combinatorics, and number theory. He collaborated with hundreds and hundreds of individuals on mathematical papers and was happy to talk with anyone as long as it was about mathematics. He referred to people who had ceased to do maths as having 'died' and so were presumably no longer worth talking to. He eschewed sex, alcohol, and music but indulged heavily in coffee and amphetamines. He did manage to win a bet by giving up drugs for a month but claimed that his self-denial served only to set the cause of mathematics back four weeks—without his stimulants blank sheets of paper remained blank. Although I never met the great man, I am left with the distinct impression that despite no one being able to put up with him for very long, he was personally perceptive and had the ability to bring out the best in people when it came to engaging their natural mathematical capacity.

The network of mathematical collaboration based around the node of Erdös has received a lot of attention. The length of the shortest chain of published collaboration (if there is one) from Erdös to a particular mathematician is known as their *Erdös number*. Erdös's own number uniquely is 0 while his collaborators have the privilege of an Erdös number of 1. My own Erdös number, I believe, is 3 (one of my co-authors claims to be a '2' and I have taken his word for it) and it is said that 90 per cent of the world's mathematicians have an Erdös number no more than 8.[1] Some, of course, have no number at all. If your collaboration chains do not reach the component containing the Erdös node then you have no Erdös number. However, one could arise even after your own death through new edges of collaboration arising from those people with whom you have jointly published.

Although Erdös, has now 'left', as he liked to put it, it is still possible to reduce one's Erdös number to as low as 2 by writing a joint paper with one of Erdös's original collaborators. However, as the years pass, and more people 'leave' it will become progressively

---

[1] During the writing of this book I have been able to verify my E. number using the Erdös Number Project website: <http://www.oakland.edu/enp/>, which is dedicated to studying mathematical research collaboration. The number of people with an E. number of 2 is nearly 7,000 and, as explained above, it is still possible to join this club.

harder to lower an existing value. As the generations pass, the centre of the Erdös network will ossify, with no new edges coming from the nodes that have 'left'. It will still be possible for newly born mathematicians to join the Erdös network through surviving members, a measure of the enduring influence of this one hub. However, the new chains must inevitably get longer and longer as time passes and whether the outer part of the network will continue to grow and flourish indefinitely as the inner realm passes into history remains to be seen.

There have of course been other prolific and sociable mathematicians down the years. Leonhard Euler, for example, although living his life in the eighteenth century, managed to produce more pages of published mathematical research than the single-minded Erdös and was arguably wider in scope. Some historical mathematical hubs were not themselves important mathematicians. Plato was such an example in the fifth century BC and in the seventeenth century a key channel of communication was through the monk Marin Mersenne. Although making only modest personal contributions, Mersenne was in contact with the leading mathematical figures in Europe and thereby disseminated discoveries throughout the community in the days before mathematical journals. His contact with the prolific but reclusive amateur mathematician Pierre Fermat was of particular significance.

The most important qualitative feature of the network of personal familiarity is the presence of the hubs, as they go a long way to giving this network its character. These hubs have very large numbers of disparate acquaintances. And a hub does not need to be an especially famous person like the Pope or the President or Erdös. There are many thousands of individuals who, in the course of their life, become well known to thousands of others. Most of us will be directly acquainted with one or two of these hubs who are likely to have very short acquaintance chains to millions upon millions of people. That is why it is a small world.

We may be tempted to think these considerations apply to any big network based on human contact such as the World Wide Web. Broadly speaking, this is the case, however six degrees of separation is not always enough. Albert-Laszlo Barabasi in his influential book

Linked: The New Science of Networks explains that experiment has shown that the degree of separation of Web sites averages about nine clicks with the largest separation rarely exceeding nineteen links. It seems that currently the Web, a nebulous object that no one can see in its entirety, is a little different from the network of human acquaintanceship. On the one hand it is run by and for people, there is a similar number of nodes in the network and, like the human species, it has spread across most of the globe. What is more, it features many large hubs: popular web pages linked to myriads of others. On the other hand, the web is a *directed* network with links often only going in one direction. Individual web pages and blogs can be lonely and seldom visited. There may be many links out of a personal web page, but few going in.

The Web can offer the illusion of allowing anyone to be part of the worldwide flow and be a real and independent player in the modern era centred around the Internet. The reality may be that the vast majority of web pages and the people behind them are invisible, even to search engines. The Web does allow similarly inclined people to communicate with one another and so form 'local' friendships even though these kindred spirits may not be geographically close and indeed may never live to shake one another by the hand. Like the rest of the human world, however, the Web is somewhat dominated by big shots and large institutions with most individuals struggling to be heard.

All the same, non-commercial networks can flourish on the web and the common thread of the participants can be tiddlywinks or terrorism, pornography or peace, death or dating. There is more freedom for individuals on the net than would normally be allowed even within very liberal democracies. No one is calling the shots and there are few guardian angels.

A key point made convincingly by Barabasi is that links in important networks are not random but the degree of their nodes follow *power laws*.* A lot of the early mathematical work on networks regarded them as 'random graphs' where the set of nodes was given and fixed while links sprang up between them by chance. However, in the real world, the networks we meet evolve. New nodes are born and begin to acquire links while old ones sometimes die along with the links binding them to the network.

Direct mapping and measurement of various important networks have verified that the distribution of the degree of nodes follows power laws as opposed to the exponential drop off that you would meet if the connections were more random. In a random network, although the degrees of nodes would vary, there would be very few nodes that were much larger than average and virtually no real hubs: nodes of very large degree. In contrast, if the degree distribution of a network follows a power law things looks very different. It is still true that the vast majority of nodes will be of low degree but there will be a reasonable proportion of nodes that are considerably larger than most. Moreover there will be a small percentage of very large nodes indeed. In a large network a small percentage can still represent a large number, for example 0.5 per cent of one million is 500. To top things off, a handful of these big nodes will be monsters. These big nodes can be very dominant, driving the development of the network which itself may be vulnerable to collapse if some of them fail. This seems to be the kind of structure we come across when dealing with most evolving, real-world networks.

Barabasi and others have studied a multitude of networks such as the Web, networks of business and personal contacts, the spread of epidemics, fashion trends and ideas, and chemical and biological interactions. In each case they have found similar qualitative features throughout. More importantly, they can model these networks mathematically, allowing prediction of their behaviour and understanding of their strengths and vulnerabilities.[2]

It is these weaknesses that are often invisible to the participants of a network who can all be left baffled when it suffers collapse or massive disruption. Examples are given varying from internet chaos to international banking and financial collapses. Events such as the Asian financial crisis of the later 1990s take everyone by surprise. In these crises it becomes apparent that there are no true experts available, for no one had a clear picture at the time as to what was happening and how far the crisis would spread. After the fact, it is

---

[2] The extent to which certain suggested mechanisms for producing scale free networks really apply to the internet and other large nets is by no means a settled question: there is an interesting referenced essay on the topic of scale-free networks to be found (at the time of writing) on Wikipedia.

possible to list the train of events and explain how each triggered the next, which gives the illusion that, with hindsight, the process is comprehensible and perhaps could even have been anticipated. Authorities try to reassure everyone that all will be sorted out and that 'lessons will be learned' but that is easier said than done. After an earthquake, it might be clear how the collapse of one building undermined another. That may allow you to build more earthquake-proof structures in future but it does not mean that you can predict when the next quake is coming or how severe it will prove to be. Although humans are the ones who create financial systems and markets it does not guarantee that the behaviour of these networks is predictable or even understandable—at least experience warns that we should not be over confident. The case has been made that networks merit a thorough investigation in their own right.

The sciences are often criticized for being too relentlessly analytical. Much research goes into breaking the object of study into constituent pieces that can be thoroughly examined and understood while too little effort is spent in understanding the overall picture. Scientists themselves have been conscious of this but, unfortunately, holistic approaches often yield little that is new. I once spoke to a biologist who said he felt sympathy for young researchers but would cringe when he heard them say they wanted to study an organism 'as a whole'. He would try to dissuade them from this mindset because, 'I just know they are going to fail'. What Barabasi and others are succeeding in doing is offering some hope for those who yearn for a holistic approach. Through the study of networks, large systems can be investigated in their entirety and genuinely surprising and critical conclusions can be demonstrated. It is obviously important to be able to tell when a network that looks stable and robust in fact is not. A great deal of interdependency can be a source both of strength and of weakness. At the same time, important networks have a life of their own. Not only are they under the control of no one but they are not even visible in their entirety. Real networks grow and evolve of their own accord, with no one person or institution having overall responsibility or even knowledge of what is going on. The problems this throws up are interesting and important indeed.

Let us return, however, to where the subject all began, a small Prussian city in the eighteenth century.

The bridges of Königsberg

The very first problem of network theory was solved in 1735 by Leonhard Euler (pronounced 'Oiler'). It was a simple question about traversing a bridge network, a problem type that has become a staple of puzzle books and mathematical riddle makers ever since. The old Prussian town of Königsberg (now Kaliningrad in Russia) lies on the banks of the Pregel river. It is known as the birthplace of the nineteenth-century physicist Gustav Kirchhoff, who has been mentioned before and will feature again later. However, Königsberg's most famous son is the great eighteenth-century philosopher Emmanual Kant who, we are told, spent his entire life in the town, never venturing more than a few miles outside it and, according to one biographer, 'never saw a mountain'. It appears that Kant led a monkish life entirely devoted to his philosophical musings. He was however fond of walks so he would undoubtedly have been well acquainted with the seven bridges that serviced the town and gave access to the banks and to a pair of islands that nestle in the river, something like what we see in Figure 3.1. The question that was asked was:

Can a person walk all the bridges of Königsberg once and only once?

The frustrated citizens brought the problem to the attention of Euler who saw that, although simple, it was unlike any problem that mathematicians had hitherto tackled. It required a fresh approach. He showed how to solve this vexing puzzle and any similar problem.

The key is to spot that what we have on our hands is really a question about the underlying network, and so to make progress, we should identify and draw that network. With respect to the network of seven bridges, there are only four places a walker can be, as indicated in the next diagram. Our first simplification in the way we look at the problem is to represent these four places (the two

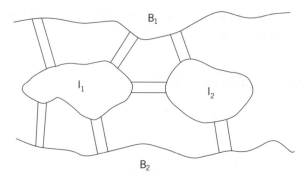

Figure 3.1 The seven bridges of Königsberg

banks of the river and its pair of islands) as nodes. Naturally we draw one line between pairs of nodes corresponding to each of the Königsberg bridges. This network distils all the information about the bridge network relevant to the question at hand (Fig. 3.2).

Euler explained that the network did not allow you to walk all the bridges just the once, and it is all to do with even and odd numbers. Suppose, said Euler, that there were such a walk that traversed all seven bridges exactly once. It would begin at some node, end at another (although this conceivably might be the same node), but there would be at least two nodes that were neither at the end nor the beginning of your walk. Focus on one of them—let us call it node X for the time being. Since X is at neither end of our walk, we

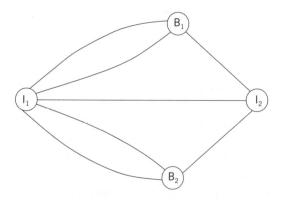

Figure 3.2 Network for the Königsberg bridges

would visit X a certain number of times, and leave X *an equal number of times*. This would make use of an *even number* of bridges overall— each time we arrived and left X, we would use a pair of bridges that we would be forbidden to use again. It follows that X must have an even number of bridges servicing it in total. Unfortunately, however, this is true of none of the nodes in our picture: I_1 is connected to five bridges, while the other nodes each have three. This shows that there is no walk with the properties we were after. It cannot be done.

This may be disappointing, but mathematics is often very good for this purpose—showing conclusively that we are wasting our time trying to find something that does not exist because it cannot: if something is mathematically impossible, then it truly is impossible. The network that represents the Königsberg bridges is certainly not a tree. It is connected but there are cycles in the network and it even displays *multiple edges* in that some pairs of nodes have more than one edge running between them (I_1 and B_1 for example have two). Nonetheless, Euler showed exactly when you can and cannot traverse a network like this. He gave a simple rule to decide the question and, what is more, he explained how you can go about finding a successful walk in the cases where it is possible. The method also allows you to decide if you can organize things so that you will return to your starting point and whether or not you can almost find the walk you need. (It is not hard to convince yourself that you can walk any set of *six* of the bridges of Königsberg but, as we have seen, never the seventh.)

Surprisingly, these claims are all quite easily explained as the basic principle amounts only to taking the idea of this example and placing it in the general setting. Before we go further in that direction however, we look at the next question, which looks different, but really isn't:

Can you walk through all the doors of the house just once?

The straight lines in the picture of Figure 3.3 represents the walls of a house with the gaps indicating doors. The challenge is to move about the house by using every door exactly the once. The diagram shows my near successful attempt where I have just missed one

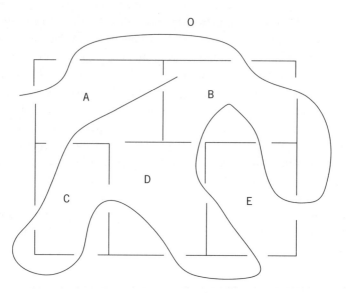

Figure 3.3 Wending through all the doors of a house

internal door. Unfortunately, neither end of my walk can reach the door I overlooked without passing through a door that I already have used. You may be tempted to wrestle with the problem yourself but you will fare no better. Underlying the problem is a network, which although a little more complicated than that of the Königsberg bridges, has the same awkward feature that is going to thwart our every attempt.

In this problem the network will have six nodes: one for each room and another for the outside of the house. The outside has no special status as regards our problem and indeed this is a facet that we have already seen in the Königsberg problem for two of the nodes represented the islands in the Pregel but the other two represented the river banks, which were regions that were effectively boundless in the direction away from the river. That, however, is mathematically irrelevant—it is only the connections that matter!

The rooms are the nodes and the doors are the edges that allow us to travel between nodes. To draw the network we just place a node in each room and one more outside and draw one edge between nodes

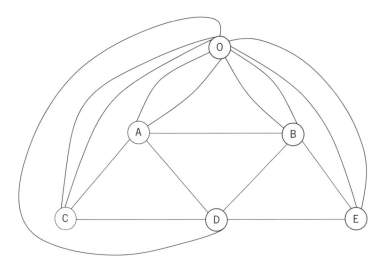

Figure 3.4 The network of our house

if there is a door between them. Having done that we recover the network shown in Figure 3.4.

Can we apply Euler's reasoning to this network also to explain our failings? If we read through the argument that worked for Königsberg, we find that it does not apply immediately, so let us take stock of our position. First, let us note the degree of each node: O is of degree 9, corresponding to the nine outside doors, nodes A, B, and D are of degree 5, while the remaining two nodes have degree 4. If we examine the argument used in Königsberg we see that there is one conclusion that will always apply: if there is a traversing walk, that is one that uses every edge (i.e. door) exactly once, then the degree of every node, with the possible exception of the first and last, must be an even number, as we leave non-terminal nodes as many times as we arrive. This is the general principle that lies at the heart of the matter: *to be able to traverse the network there cannot be more than two odd nodes.* Unlike the Königsberg Bridge Network, the House Network does have some even nodes (that is, nodes of even degree) but it does have more than two odd nodes (four in fact, O, A, B, and D) and therein lies the rub, we can't pass through all the doors just the once.

In a later chapter we shall return to Euler's Principle and consider its consequences more carefully, that being that a connected network can be traversed in such a way that you may return to your starting node exactly when the network has no odd nodes; if the network has two odd nodes, it can be traversed but the beginning and end of your path must be these two odd nodes. What if we have just one odd node? Well, that is an impossibility by virtue of the following principle.

Hand-shaking and its consequences

In relation to the general principle that sprang from the Königsberg bridges, we make a pair of related observations about general networks. If you look at any of the diagrams of networks in the book and add together the degrees of all the nodes the answer is always an even number. For example, take the tree representing the propane molecule in Figure 1.3: summing all the degrees gives $(1 + 1 + 1 + 1 + 1 + 1 + 1 + 1) + (4 + 4 + 4) = 8 + 12 = 20$, while for the network of the Königsberg Bridges (Figure 3.2) we get the sum $5 + 3 + 3 + 3 = 14$. Can you see why this is unavoidable? It is all because of the edges: each edge is incident with *two* nodes and so contributes *two* to the overall sum of the degrees, once for each end. It follows that the sum of the degrees is equal to twice the number of edges in the network and therefore must be an even number. This fact is often referred to as *The Hand-Shaking Lemma* since it implies that if several people shake hands, the total number of hands shaken must be even, as each handshake involves a pair of hands. It is one of the very few results that applies to all networks and it is convenient to have a memorable name for this simple rule to refer to it by. This lemma may seem obvious, at least once it is pointed out to you, and of little value as there would seem to be little interest in the sum of the degrees of a network in any case. However the lemma has one consequence well worth noting.

The number of odd nodes in a network is itself an even number.

Or put differently, a network cannot have an odd number of odd vertices, for suppose that it did. If we were to sum the degrees of all the odd vertices in this instance we would have the sum of an odd number of odd numbers, let's call it O, which is always itself odd. The sum of the degrees of the even vertices will be an even number, E say, as the sum of any group of even numbers is even. The total degree sum would then be $O + E$, and since an odd plus an even is odd, this would give a network where the sum of the degrees was odd, contrary to the Hand-Shaking Lemma. This then is impossible and so the number of odd nodes in a network is always an even number. For example, in that of Figure 3.2 there are 4 odd nodes, in Figure 3.5 there are 6, while in Figure 3.9 there are none at all. This corollary of the Hand-Shaking Lemma is indeed worth appreciating as it represents a constraint that applies to all networks—the network may consist of several pieces, it may have multiple edges and loops, it will make no difference: you just cannot have an odd number of odd nodes, a fact that asserts itself from time to time in the everyday world of human affairs and is not to be denied!

For example, it is not possible to have a gathering of nine people where each of them is acquainted with precisely five others in the group for if we represented this as a network of nine nodes in the obvious way, each node would be odd (of degree five) and so we would have a network with an odd number of odd vertices, and that can never happen. Alan Tucker noted in his text *Applied Combinatorics* that this very point was overlooked under the old National Football League schedule guidelines in America. At one time there were twenty-six teams divided into two equal Conferences, the AFC and the NFC. Each team played fourteen regular games each season and the recommendation was that each team should play eleven times within its own Conference and three times against teams from the opposite Conference. Can you spot the trouble with this? Just focus on the network of games within either of the two Conferences. If this guideline were somehow satisfied the scheduling network within a Conference would consist of thirteen nodes (one for each team) each of degree eleven, as we, quite naturally, draw an edge between teams that play one another. This would yield a network with an odd number of odd nodes—don't waste time getting your computer buffs to search through all possible schedules—it just cannot be done!

All this allows us to say more about our previous two examples. If any one of the Königsberg bridges were dismantled, then the resulting network would have exactly two odd vertices, those being the pair that were not joined by the bridge. This means that if one bridge were taken away, you could walk the remaining six, provided that you start and end at the points that were *not* connected by the bridge you removed. Another way of putting this is that although the seven bridges of Königsberg cannot be traversed, any set of six of them can be managed.

The story with the house is similar but not quite so simple. The two lower outside rooms represent even nodes in our network, and so it will not help you to stop up any door of rooms C or E, as that will not reduce the number of odd nodes in the network. On the other hand, if you remove an edge connecting a pair of odd nodes, that will leave the network with only two odd nodes and so it will be possible to traverse it provided that you begin at one of the two remaining odd nodes and finish at the other. For example, my near successful attempt would work if we closed off the door connecting rooms A and D, as the original picture shows. Similarly if we walled up one of the outside doors of B, this would leave B and O as even nodes and the problem could now be solved provided that the path ran between the two remaining odd nodes, A and D. (Try for yourself.) However, removing any door of C will still leave you with an impossible problem.

The next network problem, although superficially similar to our previous examples, represents a fundamentally different kind of question. We have three houses, A, B, C to be joined to the three services: Gas, Electricity, and Water. We would like to do it without any of the connecting links crossing one another.

Can the three services be joined to the three houses without the connecting lines crossing?

My first attempt (Figure 3.5) in which I have simply drawn a straight connection between each house and each service does not look too

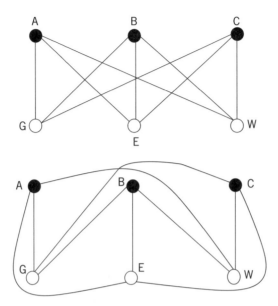

Figure 3.5 Connecting three homes to three services

promising as it has created six points where lines meet in pairs and one point where all three cross. However, it is not too difficult to do much better than that as shown in the second figure.

This attempt is an improvement, yet it is still a failure as the Water connection of *A* crosses the Gas connection of *C*. This is however the best anyone can do—it a very fundamental fact that this particular network is not *planar*, that is to say cannot be drawn on a flat sheet of paper without one pair of edges meeting.

To convince you of this, it is perhaps better to draw it another way. This network is certainly not a tree as it contains cycles: for instance a cycle of length six is given by $A \to G \to B \to E \to C \to W \to A$. If we draw the network starting with this six-cycle we draw a closed curve of some kind in the plane containing these six nodes. This leads to quite a different picture of the same network (Figure 3.6).

Although it is a strikingly different picture, we can at the same time recognize it as the same network, as the connections between the named nodes are precisely the same. With the network drawn in this fashion, the difficulty in trying to avoid edges crossing can

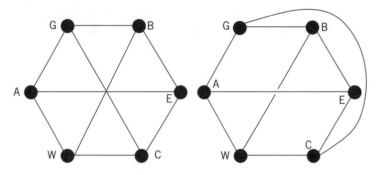

Figure 3.6 New configuration of network

be more readily seen. Let us suppose that we somehow do draw the network while avoiding the crossing of edges. However we go about it, we shall have the above cycle creating a closed curve that has an inside and an outside. Each of the three remaining edges will need to lie entirely inside or entirely outside the cycle. This will mean that at least two of these additional edges will be both inside or both outside that curve. Once one edge, such as AE, is drawn inside, however, it splits the interior of the curve into two parts with the edge forming a barrier for the two remaining edges GC and BW to cross. One of these edges can be placed outside, by symmetry it matters not which one, but let us say it is GC. However it is done, one node of the remaining edge BW will find itself surrounded by a closed curve made up of edges already drawn and cut off from the node at the other end of the remaining edge to be included (in the diagram the curve in question is determined by the cycle $GAECG$). The best we can manage is one edge crossing as we have seen before and which is repeated for this formulation in the diagram on the right in Figure 3.6.

It turns out that there are two fundamental examples of small networks that are not planar, one of which is that above that mathematicians call $K_{3,3}$ as it consist of two sets of three nodes, with each node of the first set joined to the second. The second problematic fellow is K_5, which is the so-called *complete* network on five nodes, meaning that each node is connected to the other four. This one can also be drawn with just the one unwanted crossing but that is the best you will manage—K_5, like $K_{3,3}$ is also not planar. These fellows are sometimes referred to as the 'minimal criminals' for it turns out that a network is planar unless it contains one of these two embedded inside itself in a certain fashion that will have to wait till later to be explained more precisely.

Cycles that take you on a tour

As a rule of thumb, we can say that a network will be planar unless it has too many edges compared with its number of nodes. This is indeed a very rough and ready way of putting it but it is pretty obvious that the more edges we insist the network has, the more difficult it becomes to draw them all without the edges crossing somewhere. Although having many edges is a property that is bad for planarity, it is good for another aspect that is nice to have in a network, which is that of having a Hamilton cycle. By a cycle we mean a path that begins and ends at the same node without going through any node more than once.

The idea of a Hamilton cycle is named after the famous Irish mathematician Sir William Rowan Hamilton (1805–65) and it means a cycle that takes in all of the nodes in the network. In this way it is a kind of dual idea to that of an Eulerian circuit we saw above: an Euler circuit has to pass through every *edge* of the network exactly once while a *Hamilton cycle* must do the same but for every *node*.

You will notice that we have to start being careful with our words: we are calling it an Eulerian *circuit* because it is not necessarily a cycle—a circuit can visit the same node more than once but a cycle cannot. Some books use the term *simple circuit* to mean cycle for just this reason. Another part of the definition of an Euler circuit we have not mentioned up till this point is that the circuit visits every node. This is not quite automatic: for example, if a network consisted of two components, one of which was an isolated node, a circuit could traverse every edge without visiting that node. In practice, we are only interested in connected networks when discussing Euler or Hamilton paths and so this nicety need not presently concern us.

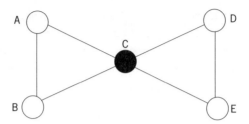

Figure 3.7 The bow-tie network

All the networks we have seen so far, with the exception of trees that of course have no non-trivial circuits of any kind, large or small, have a Hamilton cycle and they are easily spotted. In our $K_{3,3}$ network for instance the cycle we drew attention to that contributed to its non-planarity was Hamilton: $A \to G \to B \to E \to C \to W \to A$. An example of a network in which every node lies in a cycle but is not Hamiltonian is the bow-tie network of Figure 3.7.

The bow-tie has an Euler circuit $A \to B \to C \to D \to E \to C \to A$ but this circuit is not a cycle as the node C is repeated. (It *appears* that A is visited twice as well but it is not, this is an illusion due to choosing A as the starting point of our circuit—the difference with C is that we both arrive at and leave node C twice.) In order for a network to have a Hamilton cycle it cannot have a node like C. What is different about C? The special property it has is that if it were removed (along with the edges that are incident with it), the network would split into separated parts. Any circuit in the network that took in every node would have to visit such a node more than once as it passes from one of these components to another and back again. For that reason, the circuit could not be simply a cycle, so no Hamilton cycle is to be found.

Euler discovered in the eighteenth century exactly when a network has an Eulerian circuit and when it does not. No one has come up with a similar answer as regards the Hamiltonian property. There are, however, some sufficient conditions that are simple to state and guarantee that a network is Hamiltonian. For example, if each node in the network is adjacent to at least half the nodes in the network then a Hamilton cycle is an inevitable consequence. (There is one

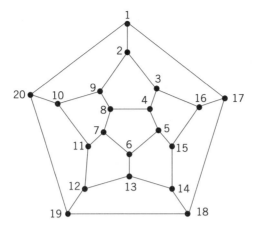

Figure 3.8 Hamilton cycle for the network of the dodecahedron

little exception, namely the network O—O.) However, in general, the question is not easy to decide, even for fairly simple networks as seen in the next example (Figure 3.8), due to Hamilton himself.

This network is called *Platonic* as it arises as the network of vertices and edges of one of the five regular Platonic solids, which are the tetrahedron, the cube, the octahedron, the icosahedron (made up of twenty equilateral triangles) and the dodecahedron that consists of twelve regular pentagons pasted together. If we were to project the shadow of a dodecahedron onto a flat sheet of paper, we could obtain the network in the diagram of Figure 3.8. In 1859 Hamilton exploited his discovery of a cycle that spanned all the nodes to invent a game based on the idea of a grand tour of twenty great cities of the world. Perhaps realizing the limitations of this discovery as the focus of a game, he sold the patent on.

The idea of devising ways of efficiently visiting a number of designated places and returning home became the *Travelling Salesman Problem*, which is still unsolved and receives a great deal of practical attention to this day. Here the call is not just for any Hamiltonian cycle but for one that minimizes length, or cost, or time of travel. That is to say the edges of the network carry *weights*, and the problem calls upon us to take these weights into consideration while finding a solution. A great many problems in economics, and in *operations research* as the associated mathematical field is known, are concerned with problems of this kind.

The numbering of the nodes in Figure 3.8 for Hamilton's original nineteenth-century problem provides a Hamilton cycle for all to see. At the same time, *finding* a cycle in the first place is not so easy and indeed we have no way to tell in advance that there is one at all. There are, however, practical approaches to conducting the search for Hamilton cycles. These come down to judicious application of the following rules:

1. If a node has degree 2 then both of its edges must be part of any Hamilton cycle.
2. No cycle not containing all the nodes can arise when building a Hamilton cycle.
3. Once a Hamilton cycle under construction has passed through a node then all of the unused edges incident with the node can be dropped from consideration.

Party problems

In our next pair of questions, we return to the subject of networks of friendship and acquaintance, but we begin by examining the properties they possess, even on a very small scale. The Ramsey question seems so innocent and simple yet represents the tip of an enormous mathematical iceberg, that of Ramsey Theory. The problem is this: how many people do you need at a party to ensure that there will be a triangle of mutual acquaintances, or a triangle of three mutual strangers?

The answer has to be more than five, because of the following dinner party arrangement. Imagine five people sitting down to their meal in such a way that each person knows the two people sitting next to them but not the two others. This is certainly possible: if we sit the five around the table and let them join hands with the two they know we obtain a simple five-cycle as their network of acquaintantship (Figure 3.9).

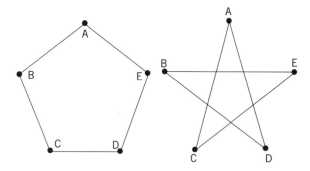

Figure 3.9 Acquaintanceship of five dinner guests

This network has no triangles, as no three people are all mutually acquainted. Neither is there a triangle of strangers: for example, look at person A (by symmetry, it matters not which person we focus on); the guests that A does *not* know are C and D but they are mutual acquaintances.

We can display the network of strangers but we get what at first sight looks a bit of a tangle. Indeed a slightly sinister pentagram appears as the network of strangers. However, in this case, these two networks are in reality the same: if we list the nodes of the network on the right in Figure 3.9 in the order $A \rightarrow C \rightarrow E \rightarrow B \rightarrow D \rightarrow A$, we see that this pentagram is also just a simple five-cycle, identical to the original—in particular, it has no triangle of three mutually connected nodes.

In general, the network we obtain when we take the same set of nodes, delete all the edges, and then connect the pairs of nodes that were previously not connected is called the *complementary network* (an idea that only makes sense when discussing networks that do not have loops nor multiple edges running between pairs of nodes). The five-cycle is an example of a *self-complementary* network, as the complement is also a five-cycle, as you can see upon closer inspection.

The answer to our Ramsey Problem is therefore at least 6 and if you play around with networks representing six or more people long enough you will convince yourself that 6 is indeed the answer—but how can we be sure?

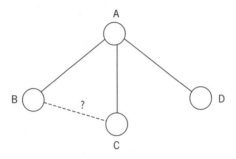

Figure 3.10 Acquaintances at a party

The difficulty is that, given six people, there are many possible arrangements of acquaintanceship that could arise between them. Our argument has to be able to cope with them all. If we go about it the wrong way, we will soon be lost in a multitude of cases. An effective and simple argument however is to hand but takes a little sharp observation.

Consider any six people at the gathering and focus on one of them, called *A* (Figure 3.10). Of the other five, either *A* knows at least three of them or, if not, there are at least three he does not know. (This is the one place in the argument where we use the fact that there are at least six people present.) Let us suppose for the moment that three of the people are known by *A*. Then either these three people have not met before, in which case we have found a required triangle of mutual strangers, or at least two of them, let us call them *B* and *C*, do happen to know each other. But then the threesome of *A*, *B*, and *C* form a trio of mutual acquaintances. The argument is now essentially complete as the alternative case in which none of *B*, *C*, or *D* is known by *A* is the same—either *BCD* is a triangle of mutual acquaintances or if not, one pair of them, together with *A*, form a triangle of strangers. We conclude that whenever six or more people gather together, there is either a triangle of mutual acquaintances, or a triangle of strangers (and perhaps both).

We have shown that 6 is the smallest number of nodes that a simple network must have in order to *guarantee* that either the network or its complement has a triangle. This kind of result will always stimulate a response from someone with mathematical training for generalization looks a real possibility. The query that naturally comes to mind is: How large a network do we need to ensure that either it or its complement has a *clique* of four nodes that are all mutually connected? And generally, how big does the network have to be to guarantee that a clique of a given size k is present in either the network or its complement?

These are very good and very tough questions—indeed no one knows the answer to the latter, not even for $k = 5$. However, we do know that there *is* an answer, for even that is by no means obvious. After all, it is conceivable that it might be possible to arrange a party of a size exceeding any given number in which there was no group of four friends and no group of four strangers. However it is known that once we have 18 or more people, this becomes impossible—we say the fourth *Ramsey number* is 18. What the English mathematician and economist F. P. Ramsey (1903–30) proved in the 1930s was that Ramsey numbers always exist—for any k there *is* a minimum number n (the size of n depending on k), such that any party of n or more people has a clique of k mutual acquaintances or a k-clique of mutual strangers. However, the exact values of these Ramsey numbers generally remain a mystery, but they do exist— Ramsey proved it and a demonstration can be found in the final chapter.*

The next question on networks is also most easily appreciated in terms of a party.

At a party, must two people share the same number of friends?

You may not have realized that this is the case but if you experiment with any gathering, large or small, real or imaginary, it will always turn out to be true. Why should that be? Once again, it is to do with the nature of networks that continues to spring surprises. The demonstration, however, involves a fundamental fact about counting often called the *Pigeonhole Principle*. This extremely important observation is little more than a piece of mathematical common sense: if you have more letters than you have mail slots (or pigeonholes as they are sometimes known) to place them in, then at least one slot must have at least two letters. This observation, simple though it may seem, is extremely important and arises time and again in *combinatorics*, the mathematics of counting, to yield conclusions of inevitability.

For example, in a town of 400 souls, at least two have the same birthday as there are more people than possible birthdays. Indeed we can say more: even allowing for the 29th of February, there must be at least $400 - 366 = 34$ people in the town who share a birthday with some other citizen because the number of people for which this is *false* cannot exceed 366, the number of birthdays available. No one may have any idea who these 34 people are (and there could of course be more than that) but it is a mathematical certainty that they are there!

I must confess that we already slipped in versions of the Pigeonhole Principle in the argument about the Ramsey Problem and even earlier when we were showing that the gas, electicity, water configuration was not planar. Here we made the simple observation that if we draw three edges connecting points of a cycle, then at least two of them have to be either inside or outside the cycle: this corresponds to three letters slotting into two pigeonholes. In the Ramsey argument recall that we focused on A, one of the six people, and divided the other five guests into two types, those acquainted with A and those that were not. In doing this we are in effect putting five objects into two slots and concluding that one slot must contain at least three. The general principle at work here is the following: if we put *more than* $m \times n$ objects into m slots then at least one slot must have *more than* n of them. In the Ramsey Problem $m = n = 2$, so that $m \times n = 4$ and we are placing the 5 people into 2 categories, so that at least 3 are of the one type. There is nothing difficult in any of this but I draw your attention to it just to emphasize how often this trick comes up in reasoning of this kind.

There are many clever exploitations of the principle that have been devised to prove surprising results on inevitability within large collections. A simple example comes from considering the set of all numbers up to the nth even number, $1, 2, \ldots, 2n$. If we now take *any* set of $n + 1$ of these numbers, at least two of them will have no common factor. This follows at once from the fact that two numbers from your collection must differ by only 1 as it is plainly impossible for the gaps between all the numbers in your collection to be more than 1, for then the largest member of your set would be at least as great as $1 + 2n$. (That is to say, to have at least 2 letters in n slots requires at least $2n$ letters.) For example, for any collection of six numbers from $1, 2, \ldots, 10$, two of them must be consecutive integers. Any factor of the first of these neighbouring numbers will leave a remainder of 1 when divided into the second so there are your two numbers with no common factor (other than 1). This is not very surprising, although we should add that the observation cannot be pushed any further for it is easy to find a set of n numbers in this range, all with a common factor of 2, namely all the even numbers, $2, 4, \ldots, 2n$.

A more surprising observation about this set of $n + 1$ numbers is that one of them must always be a multiple of one of the others. This is proved through a rather more deft application of the Pigeonhole Principle.*

This Pigeonhole idea also surfaces in our party question as I will now explain. Suppose that there are n people at the party, where n must be at least 2, for otherwise we would have no party. The most friends one of the party goers can have at the event is $n - 1$: for instance the girl hosting the party might have only invited her own friends. The least number is 0. This sounds sad but is possible: the party might have an unwelcome gatecrasher. Bear in mind then that every individual at the party has a *friend number*, the number of friends at the party, and this number lies in the range 0 to $n - 1$ inclusive.

Now suppose, contrary to what we are expecting, no two people at the party did have the same number of friends, that is to say, everyone's friend number is different to everyone else's. This is not easy, but looks just possible: there are n different numbers distributed among the n people, which means that each of the possible friend numbers $0, 1, 2, \ldots, n - 1$ is taken up exactly once. There is, however, one final twist that renders this impossible. Some person P scores 0 (no friends) while some other, Q say, scores the maximum $n - 1$. This means however that Q regards everyone else at the party as her friend, *including the otherwise friendless P*. However, if P and

Q are friends, then P cannot score 0 after all. We have thus found a genuine incompatibility arising from the assumption that everyone has a different number of friends, and so this cannot be the case. Therefore there are always at least two people at the party with the same number of friends, and this goes for any gathering anywhere, anytime.

Of course, all this is really telling us something about networks, at least about any network that could represent party acquaintance, and these are indeed pretty general. There are no real restrictions on a 'party network' except that the number of edges between any pair of nodes is either 0 or 1—no multiple edges like those we saw in Königsberg, and no edge passes from a node to itself giving us a *loop*.[3] This type of network that forbids multiple edges and loops is sometimes called a *simple network*. The party argument is really telling us that in any simple network there must always be two nodes of the same degree.

This completes our collection of problems for the moment. The next chapter introduces a simple question which has a simple answer, but it seems, no simple solution.

---

[3] Aristotle assures us that a man may be his own friend if he is a good man, so his friendship networks might have loops: however such introspection is not entertained here.

4

Colouring and Planarity

This chapter begins and ends with questions that can be resolved through talking about colouring the vertices of networks while respecting certain rules. It is striking that a topic that sounds very technical lies at the heart of a variety of questions that vary from the colouring of maps, to the guarding of museums, to deep mathematical questions that are applicable particularly in economics. The common thread throughout these investigations is planarity of the networks but its development from these problems is surprising and in one instance the application of network ideas emerges very much out of the blue.

The four-colour map problem

The most famously difficult problem in network theory is that of the four-colour map problem. Once again, at first sight, it seems not to be a network question. It is a fascinating question of a rare type: a mathematician could explain it to anyone he or she meets. After five minutes each would understand the problem perfectly and neither would be able to solve it. It does seem now, however, that an alliance of men and machines has conquered this problem. It first arose at University College London in 1852 where a mathematics student named Francis Guthrie asked himself how many colours are needed to ensure that a map may be coloured so that any two bordering countries had different colours. He soon came to the conclusion that the answer is just four but could not see how it might be proved. The problem was eventually brought to the attention of one of his lecturers, the famous logician De Morgan who, like Guthrie, could not see how to tackle it.

It took over twenty years before the problem was taken very seriously. Mathematicians, like most professionals, are busy and nothing vital seemed to be riding on this problem—in these early days, no great acclaim would be attached to the first who solved it, especially if it turned out to be quite straightforward in the end. Yet a simple problem that cannot be solved should never be ignored as little mysteries often harbour deep principles and can yield rich rewards. Part of a mathematician's skill and training is the capacity to spot something interesting and new. The four-colour problem highlights an aspect of networks that was touched on in the previous chapter, that of planarity.

To recast the four-colour problem as a question about networks, we first need to make the original question more precise by explaining exactly what we mean by one country bordering another. We do *not* mean merely sharing a common point. For example, the US states of Arizona, Utah, Colorado, and New Mexico all meet at a single point, known as the Four Corners. We do not regard the diagonally opposite states (Arizona and Colorado, New Mexico and Utah) as sharing a common border. If we did, there would be potentially no limit to the required number of colours: we could have any number of countries shaped like slices of pie, meeting at the common point in the centre and we would then need as many colours as countries to colour the map as, counting a point as a common border, each of the countries would border every other one on the map. Moreover, every country must consist of a single connected piece—that is to say it should be possible in principle to travel from any point within a country to any other point without crossing a border. Again if we allowed one country to consist of any number of pieces, these enclaves and disjoint regions would lead to maps requiring any number of colours.

Having said that, we can show how to introduce a network, indeed it will always be a planar network, associated with a given political map. We need to do this in a way so that, as with the Königsberg bridges, the network distils all the important information in the problem. The actual question posed will then have a reformulation in terms of features of the network that we can then hope to solve. This new problem should be equivalent to the old—that is to say solving one should solve the other, and this is a theme that often occurs in mathematics.

The map itself could be regarded as a planar network where the edges are the borders and the nodes are the points where borders meet. It is more enlightening, however, to construct another network, known as the *dual* of this one, when it comes to the map colouring question. To construct this sister network we represent each *country* by a node and join two nodes by an edge if the corresponding countries share a common border (as we did in the 'house' problem in passing from Fig. 3.3 to Fig. 3.4). This network then tells you whether the members of any given pair of countries are contiguous or not, which is all that matters in this question. Colouring the countries of the map then amounts to doing the same for the nodes. The question now becomes, can we colour the nodes of the network, using no more than four different colours, in such a way that two *adjacent* nodes, that is two nodes connected by an edge, always have different colours?

This is a reformulation, but it does not yet represent a complete reformulation of the problem, as not all networks arise as 'map' networks. For example, consider the network K_5 that consists of five nodes, each joined to all the others (Figure 4.1).

No less than five colours are required for K_5 because every node must be coloured differently to every other as every node is adjacent to every other. Given that Guthrie was right, and we only need four colours for any map, the network K_5 must not arise as the 'map' network of any real map. What, we might ask, would a map look like if it was associated with a K_5 network? There would need to be five regions in the plane, with each region bordering every other. If you try and draw such a map, as De Morgan did, you will find yourself frustrated! You can certainly draw four regions with each having a common border. However, when drawing the fourth one, you will cut off one region from the outside, and the fifth region you draw will not border this now isolated one. This much De Morgan

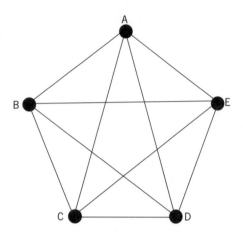

Figure 4.1 The complete network on five nodes

wrote to his friend Hamilton (of Hamiltonian cycle fame) when the problem was first disseminated.

And it is possible to show that this cannot happen. It rests on the observation, explained more carefully in a moment, that the network associated with any map is planar, that is to say it can be drawn without the edges crossing. This is an idea we first met when we considered the utilities connection problem in the previous chapter, but let us now look at it in a more thoughtful fashion.

Consider the network K_4, which consists of four nodes all joined to one another. Our first attempt to draw K_4 might result in the picture on the left in Figure 4.2 but, by redirecting one edge, we see that K_4 can be drawn without edges meeting anywhere (other than perhaps at common nodes) and so K_4 is planar.

A network that is planar may not look so, not only at first glance but even upon closer inspection. The network $K_{2,2,2}$ consists of three pairs of nodes: each node is not connected to itself or its partner, but is adjacent to each of the other four. The first picture you might draw of this network could be that of Figure 4.3, which, with its cobweb of unwanted edge crossings, looks too complicated to be untangled.

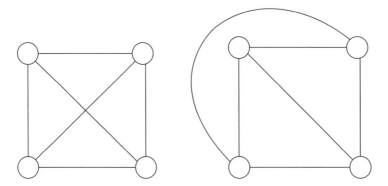

Figure 4.2 The complete network on four nodes is planar

However, the same network can be pictured in a plane fashion, without any unwanted crossing of edges, as in Figure 4.4.

We call a picture like that of Figure 4.4 a *plane network*, meaning that it is a representation of a planar network, drawn in a fashion that exhibits its planarity, with no pair of edges meeting except perhaps at their endpoints where they share a common node. There is no requirement that edges are straight lines; however, there happens to be a curious theorem that assures us that if a network is planar, then it is possible to draw a plane representation of it in which all the edges are indeed straight line segments.

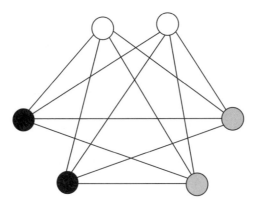

Figure 4.3 A planar network looking non-planar

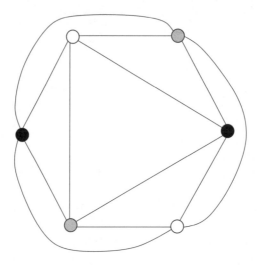

Figure 4.4 The previous network shown to be planar

Since the pictures of our networks are getting a little more complicated, this is a good place to pin down the idea of what we mean when we say two pictures represent the same network, an idea we have taken as read up until now. An example we have already looked at was that of Figure 3.9, which consisted of a five-cycle, drawn as a pentagon, and its complementary network, which was then naturally drawn as a pentagram. Since the pentagram turns out to be just another five-cycle, the two networks are essentially the same. The way we make this precise is to label the nodes of the first network, N, with the letters a, b, $c, \ldots$ and those of the second N', with the corresponding dashed letters a', b', $c', \ldots$. If the networks are really the same, there must of course be the same number of nodes in each network, but we need more than that. We also insist that whenever two nodes, u and v say, are connected by an edge, then so are u' and v' and, just as importantly, if u and v are not adjacent, nor are u' and v'. To be more precise still, we insist that *some* labelling of the nodes of N' can be found in which the number of edges running between any two nodes u and v of N is the same as the number of edges running between u' and v' in N'.

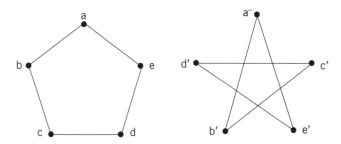

Figure 4.5 The pentagon and pentagram are the same networks

Only if that is the case do we say the networks N and N' are the same.

The word mathematicians use is that these two networks are *isomorphic* and the labelling of the nodes of N' is called an *isomorphism*. This is a technical term but it is handy to have a word for this situation when two mathematical objects are not absolutely identical yet they are the same in all the ways of current interest— it would be wrong, after all, to say the two pictures were 'equal' when there are obvious differences: it is just that the pictures do represent exactly the same arrangement of connections of the nodes. One isomorphism for the two five-cycles is then given by Figure 4.5.

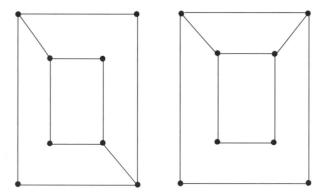

Figure 4.6 Similar but different networks

In a similar way, we can label the networks in Figures 3.5 and 3.6 to show that all those networks are the same.

For complicated networks, in general it can be a difficult problem to decide whether or not two networks are isomorphic to each other or not. If you suspect not, you need to find a network feature of one that does not occur in the other. There are the obvious things to check: do the networks have the same number of nodes and edges as one another? If they pass this first test, you can set them others. Do they have the same number of nodes of each degree? If you still have not distinguished one from the other, you need to look for more subtle differences. For example, the two networks of Figure 4.6 are genuinely different.

Each network has eight nodes and ten edges and they have equal numbers of nodes of degrees 2 and 3 (four of each). However, in the first network no node of degree 2 is adjacent to another degree 2 node but this happens for two such pairs on the right. Any isomorphic labelling of the nodes in the two networks would preserve features such as this, and so there can be no isomorphism between them—they are similar but different networks.

Returning to the network of a map, we examine how it is constructed. We place a node in each country or region as the case may be, and join two nodes if the regions share a common border. We give an example in Figure 4.7 based on the states and territories of Australia. Like the USA, Australia has its capital, Canberra, enclosed in a special Federal region, the Australian Capital Territory, an enclave within the state of New South Wales (N).

It is usual to insist on no colour clash with the outside of the region as well so we place a node in each state and one in the ocean. The network we generate by doing this is always planar for it can be drawn without edges meeting except at endpoints. This is because each edge may be drawn as a diagonal of a four-sided figure the opposite corners of which are the pair of nodes connected by the edge and the two ends of the common border associated with the regions of the edge—no other edge has call to enter this four-sided region so that no unwanted crossings need arise. We call this collection of nodes and edges the *plane network* of the map. For

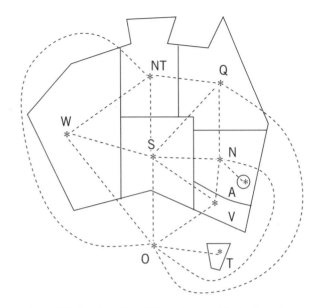

Figure 4.7 Construction of the network of Australian states

clarity, the plane network of the Australian states has been redrawn in Figure 4.8 with all edges straight, which is always possible.

Let us return now to the idea of five regions that all share a common border. If this were possible, the network of the map would be a planar network of five nodes with an edge running between each pair, giving ten edges in all. This network is called K_5, the *complete network* on five nodes. However, try as you might, you will not be able to represent this network in the required plane fashion— the best you will be able to manage is a picture along the lines of Figure 4.9, which has just one pair of edges crossing.

For suppose we somehow managed to draw K_5 in a plane way so that no edges crossed. The network will then have a cycle $A \rightarrow B \rightarrow C \rightarrow D \rightarrow E \rightarrow A$ that will form a closed figure, splitting the plane into an inside and an outside. There are still five edges, or let us call them arcs as they do not need to be straight lines, to be drawn so that at least three of them will be outside the cycle and the remainder inside, or at least three will be inside with the rest

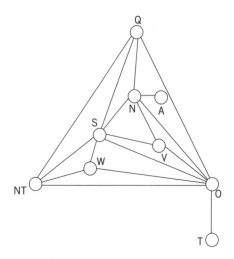

Figure 4.8 Plane network of Australian states

outside the closed cycle. However, it is not even possible to draw two of these arcs inside the figure without them crossing unless they begin at the same node, and then it is certainly not possible to draw a third inside without an unwanted crossing (try it and see!). What is more, there is no difference between the inside and the outside of the figure as regards the validity of this argument—only two of the required arcs can be drawn outside the figure, and then they must have a common endpoint. It follows that, however you go about it,

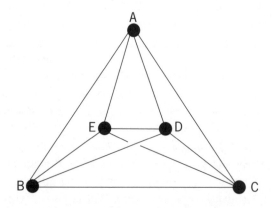

Figure 4.9 Near plane drawing of K_5

you will not be able to draw the final tenth arc without crossing an arc already drawn.

And so it is not possible to have five regions on a map any two of which share a common border. You should not imagine that this observation proves that the four-colour conjecture is true, for it only shows that, with any map, you will be able to four-colour any five regions with differing colours each side of any border, but it is conceivable that, if there were many regions to colour, interactions between sets of regions could make it impossible to use just the same four colours throughout.

Returning to the problem itself, it has had an interesting genesis all its own. Guthrie himself never published anything on his own problem, although he became a mathematician in South Africa and also contributed to botany with several plant species bearing his name. There were only a few inconclusive publications on the topic until 1879 when A. P. Kempe claimed to have solved the problem in the affirmative and on the strength of his paper was admitted to the august ranks of the Royal Society. However, 11 years later, the American mathematician P. J. Heawood identified a flaw in Kempe's proof, and one it seems that could not be patched up. Kempe's argument was, however, of some value, as Heawood pointed out that the Kempe technique could be used to show that no more than *five* colours were required for any conceivable map, but that was as far as it would take you. The *Four*-Colour Problem was again open, and the race was on to find a solution to this question that was evidently tougher and more important than a casual glance might have you believe.

Despite some progress, no solution was found until 1976, and even then, it was not of the kind that mathematicians were used to. Kenneth Appel and Wolfgang Haken verified that the Guthrie Conjecture was true. It was, however, more of a verification than a normal proof, for their approach was to split the problem up into over a thousand different cases, each of which could be checked by a direct calculation. These calculations were, however, enormous, and could only be done by computer.

This was something of a shock to the mathematical community. Was this a proof or not? Over the intervening 30 years, others have revisited this problem and no mistake has been found. In 1996, Robinson, Sanders, Seymour, and Thomas tried to verify the original proof of Appel and Haken for themselves but 'soon gave up'. Instead they devised their own program, similar to the original, and reached the same conclusion. They claim their method is much quicker, yet the proof is still very much 'computer assisted'.

The mathematical world has now grown more relaxed about 'Computer Assisted Proof', where part of the argument depends on computer verification. Perhaps certain combinatorial problems can only be solved in this way. That the Four-Colour Problem seems to be one of these is something of a surprise—it is by no means obvious that it can be reformulated in a manner that leads to this. It may yet be the case that someone will find an ordinary proof that does not involve computation. (And make no mistake, this would still be seen as an astounding achievement and the reputation of whoever came up with it would be secure!) A more reasonable hope would be that a better way of doing the necessary calculation will be found so that independent checks will be relatively easy to carry out and so the validity of the proof will be beyond all doubt. There is certainly nothing wrong with a computer assisted proof and, in principle, it is no different from one that does not involve a machine. However, as with any very long proof, the chance of a crucial mistake going undetected is high and the very existence of such a proof can inhibit people from trying to find a better one.

How edges can ruin planarity

This is something we observed in a casual way before but it is possible, with a little thought, to be much more precise and, as we shall see, precision brings with it other rewards. Any network has two numbers associated with it: n, the number of nodes, and e, the number of edges; with a planar network we can associate a third, f, the number of faces of the plane figure, a *face* being a region bounded by the edges and not containing any smaller region bounded by edges. For instance, if we take the example of the plane network of Figure 4.4, the hexagon in the diagram is *not* a face as it contains four smaller faces of the network. It is convenient to continue to count the outside of a plane network as another face. For example, in the plane copy of K_4 above (Figure 4.2), we see that $n = 4$, $e = 6$, and $f = 4$. For the network of Figure 4.4 we get $n = 6$, $e = 12$, and $f = 8$, while for the network of Australian states (Figure 4.8) we count up $n = 9$, $e = 17$, and $f = 10$. (Sometimes when counting e, you are more liable to get the count right by using the Hand-Shaking Lemma, explained in Chapter 3: sum all the nodal degrees and divide by 2!)

Clearly, however the plane network is drawn, the numbers n and e must remain the same, but this is not so clear as regards f. It *is* true, however, since for any connected plane network the three numbers are linked by a very simple equation:

$$n + f = e + 2.$$

Rewriting this to make f the subject of the formula we find that: $f = e - n + 2$. This is easily tested on all the examples we have met: for instance, for the map of Australia we verify that $9 + 10 = 17 + 2$. The reason why this relationship persists in any connected network can be seen as you draw the network, one edge at a time, adding any new nodes as they arise. You will note that when you draw the first edge we have at that stage $n = 2$, $e = 1$, and $f = 1$ (we always have the unbounded outside face) and so the formula for f is respected. At every subsequent stage, as we draw a new edge connected to the body of the network, we increase e by 1 but we either add a new node, and so increase n by 1 or, if we join two existing nodes, we split an existing face into two, so increasing f by 1. In any case, the two sides of the equation $n + f = e + 2$ remain in balance and so it continues to hold true.

This formula, familiar to Euler by 1752, can be used to show that in a planar network the number of edges e cannot exceed $3n - 6$. To reveal this very precise fact requires a little thought. If you are unused to this kind of tight reasoning and the manipulations involved, the following argument may all look a little daunting. It is not important to remember the details but at the same time it is worthwhile trying to follow it through as, if nothing else, it gives a good example of how mathematicians think things through. The desired conclusion follows from a couple of careful observations.

Before we explain why this inequality always applies, note that when studying planarity, we only need consider simple networks, that is those that lack loops or multiple edges. Given a simple plane network, it is clear that we can decorate it with as many loops and multiple edges as we wish without spoiling the planarity. For that reason, only the underlying simple network of a given network, where we strip away any loops and coalesce any multiple edges into a single edge, need concern us. Also, a network is planar if and only if each of its components is planar, so for the remainder of the discussion we take our network N to be simple and *connected*, meaning that it has but one component.

Suppose now that N is drawn in a plane fashion with the number of nodes, edges, and faces being designated by n, e, and f respectively. Count up the number of edges of every face, take the sum of all these numbers and call this total T. Now each edge lies on the boundary of at most two faces (it is possible for an edge to be surrounded by just one face—this happens for two of the edges in Figure 4.8) and so we infer that T is no more than $2e$; we write this symbolically as $T \leq 2e$. Since there are no multiple edges or loops, each face is bounded by three edges or more, meaning that each face contributes at least 3 to the sum T, and since there are f faces in all we see that $3f \leq T$; putting these two inequalities together reveals that $3f \leq 2e$ in a plane simple network. Now we know that for a plane network, $e = n + f - 2$, and multiplying through by 3 gives $3e = 3n + 3f - 6$. Since we have discovered that $2e$ is at least as large as $3f$, replacing $3f$ by $2e$ on the right hand side of the equality reveals that $3e \leq 3n + 2e - 6$; and finally, taking the number $2e$ away from both sides of the inequality we deduce that $e \leq 3n - 6$ in any planar network.

This fact immediately disqualifies the complete network on five edges, K_5 from the realm of planar networks as for K_5 we have $e = 10$ while $3n - 6 = 3 \times 5 - 6 = 15 - 6 = 9$, so that K_5 has one too many edges to be drawn in a plane fashion. However, the other minimal criminal, $K_{3,3}$ of the gas, electricity, and water problem (Figure 3.5) momentarily escapes our net as here we have $e = 9 \leq 3n - 6 = 3 \times 6 - 6 = 12$ and so our inequality rule is respected by $K_{3,3}$.

It does, however, surrender to the following argument, similar to the one that has just been put to you. Since the edges of $K_{3,3}$ always run between two sets of three nodes, it follows that any cycles in $K_{3,3}$ must be of even length. In particular, there are no triangles and so, in any plane picture of $K_{3,3}$, assuming that one is somehow possible, each face would be surrounded by at least *four* edges. This gives us a stronger statement than before when we compare f with e, namely that $4f \leq 2e$. The Euler equation, when we multiply both sides by 4, says that $4e = 4n + 4f - 8$. Replacing $4f$ by $2e$ gives the inequality $4e \leq 4n + 2e - 8$. Taking $2e$ from both sides and dividing all terms by 2 then yields the conclusion that in a plane version of $K_{3,3}$ we have $e \leq 2n - 4$. Putting $e = 9$ and $n = 6$, however, then results in the false statement that $9 \leq 8$ and so we have a contradiction. Therefore our assumption that $K_{3,3}$ *could* be drawn without edges crossing must be wrong and, like K_5, this little network is not planar. Our two minimal criminals have now been convicted by two separate strands of evidence!

Another handy fact that follows at once from our inequality is that any planar network must have a node of degree no more than 5, for if we suppose to the contrary that a planar network existed in which every node has degree at least 6, then its number of edges would be at least $\frac{6n}{2} = 3n$, which exceeds the maximum possible value of $3n - 6$. This fact allows us to prove quite easily that any map may be coloured with *five* or fewer colours.* We know now however, thanks to Appel and Hanken, that you never need more than four.

The key fact determining whether or not a network is planar was discovered by Kuratowski in 1930 for he proved that a network will always be planar *unless* it contains a copy of either K_5 or $K_{3,3}$, and so these two examples are the root of all the difficulty. This has become a model for many theorems in combinatorics which have the style that an object will enjoy a certain property unless, lurking inside it, is a copy of an object from a particular list of known culprits. The notion of 'containing a copy' is slightly more subtle than having an exact replica of either network, and indeed a little thought reveals that there must be some caveat involved. For example, consider the

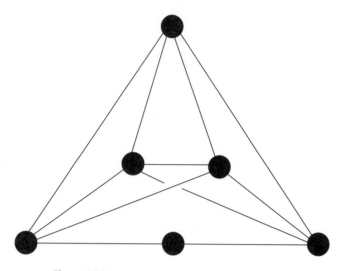

Figure 4.10 Complete network with nodes inserted

network K_5 (as pictured in Figure 4.9 for instance) and insert an extra node along one of the edges as shown in Figure 4.10.

We can see two things at once in this example. First, plonking an extra node along one edge is not going to turn this non-planar network planar. If we had a plane version of this fellow, we could just rub out the extra node and create a plane version of K_5, which we have seen is impossible (in two different ways). On the other hand, the network does not, strictly speaking, contain a copy of K_5 as there is no set of five nodes in which every node is adjacent to every other. This, however, is the extent of the complication and mathematicians have a big word for this. We say that two networks are *homeomorphic* if one can be obtained from the other by inserting or erasing 'trivial' nodes of degree two. In particular, a network that is obtained from another by introducing nodes along edges in this way is a homeomorph of the original. For example, any two cycles are homeomorphic. The precise statement of Kuratowski's Theorem is that a network is planar unless it contains a homeomorph of either K_5 or $K_{3,3}$ (in which case it is not).

An application of this idea, which does not go quite the way you might anticipate, is given by the next network (Figure 4.11).

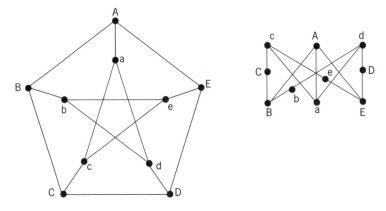

Figure 4.11 The Petersen Network is not planar

The rather sinister looking network on the left consisting of a pentagon connected to a twisted copy of itself, goes by the name of the Petersen network. It is an example of what is known as a *regular* network, that is to say a network in which every node has the same degree. The only connected regular networks of degree 2 are the cycles of any positive length. If we allow the network not to be connected then its components consist of cycles possibly including isolated loops. The only regular networks of degree 1 have components consisting of a single edge, while a regular network of degree 0 just consists of a collection of isolated nodes. The Petersen network is an instance of a *trivalent* network, that is a regular network of degree 3. Other examples of regular networks are the *Platonic networks* that arise from the regular solids such as that of the dodecahedron as seen in Figure 3.8. Any regular solid looks the same at each vertex: in particular the degree of each node of the corresponding network of connections is the same. The network of the cube is also trivalent as it consists of eight nodes all adjacent to three others. Any complete network K_n is of course regular of degree $n-1$, as every node is adjacent to all the rest.

As the Petersen network looks very reminiscent of K_5, it is not surprising that it is not planar. If we search, as would seem natural, for a homeomorph of K_5 lying within, we will be frustrated and it is not hard to see that no such thing exists; every node in the Petersen graph has degree 3, while in K_5 every node has degree 4. Adding or deleting extra nodes along the edge of a network cannot destroy or create nodes of degrees 3 or 4 and so it is not possible to find the required copy of K_5 inside the network in that way.

What is lurking within the Petersen network is a copy of $K_{3,3}$, and this is revealed by deleting some edges of the network, and finding a suitable homeomorph. We reason as follows. If the Petersen network were planar, then so would be the network that results from deleting some edges. (Quite generally, erasing some nodes and edges of a planar network will leave you with a planar network, as deleting edges will not create edge crossings that were not present before.) In particular, we drop the edges except those shown in the diagram on the right in Figure 4.11. The copy of $K_{3,3}$ then appears for all to see— the network pictured is none other than $K_{3,3}$ itself with four edges subdivided with the additional 'trivial' (that is degree 2) nodes C, b, e, and D.

There is a direct connection between the Petersen network and K_5, however, in that it is *contractible* to K_5, meaning that we can deform the network into a copy of K_5 by sliding nodes together so that the edge joining them disappears. Another way to think about contracting a network without the talk of sliding around is to delete a pair of adjacent nodes u and v and adjoin a new node w that is adjacent to all those nodes to which u or v was adjacent. This has the same net effect as 'identifying' u and v by coalescing them into a single node. The idea of contractibility does lead to another different but similar criterion for planarity of a network, that being that a network is planar unless it can be contracted to one of K_5 or $K_{3,3}$.

Rabbits out of hats

There are many problems which, although not stated in terms of networks, lead to questions that involve connected objects. The underlying networks may not be obvious at first glance sometimes because the nodes do not represent solid objects but rather processes or ideas. However, as we become more experienced in this type of question, the emergence of a network representation is not surprising and indeed becomes an expected way of coming to grips with the essential connections of the problem to hand. What is much more exciting, even stunning, are the cases where difficult questions that look to have nothing whatever to do with networks are completely solved by an application of network ideas. The most breathtaking examples often involve planarity and colourability.

So far we have only met the idea of node colouring in the context of the four-colour map problem but it is a notion that can be applied to any network. By a *colouring* of a network is usually meant a colouring of the *nodes* of the network in such a way that adjacent nodes have different colours. The smallest number of colours required is called *the chromatic number* of the network. We say a network is *n*-colourable if it can be coloured with *n* or fewer colours.

For example, any cycle is 2-colourable if the number of nodes is even, but a cycle with an odd number of nodes will need three colours, a fact that becomes clear if you examine a pair of typical examples. Indeed it can be proved that a network is 2-colourable unless it contains a cycle of odd length.* The odd cycles then play the role of the 'criminals' in this context. On the other hand the complete network on *n* nodes has a chromatic number of *n* as no node can afford to carry the same colour as any other as each pair has an edge running between them.

1. Guarding the gallery

A museum gallery needs guards stationed at fixed points to keep an eye on what is going on so that no point is allowed to remain invisible to the eyes of the guards. In 1973 Victor Klee asked what is the minimum number of guards required for a museum with *n* walls? Nowadays the guards may have been replaced by surveillance cameras but the problem is much the same. The guards are fixed in position but can swivel about, seeing freely in all directions. For that reason, quite often only one guard will be required.

An enclosed shape is called *convex* if the line joining any two points inside the shape stays entirely within it. For example, circles, rectangles, ellipses, and regular hexagons are examples of convex shapes in which case you only need one guard and he can stand

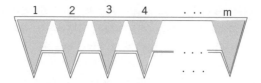

Figure 4.12 *m* guarded regions for 3*m* walls

anywhere he likes. A Christmas tree star on the other hand is not convex as a line joining two points near separate tips of the star will cross outside the figure. However, if you have a star-shaped gallery you can still get by with one guard—as long as he is posted somewhere near the centre of the star, he can see every wall in the gallery as he spins around, without moving from his spot.

On the other hand, a gallery with lots of side chapels might require a separate guard for each chapel. A particularly labour-intensive museum to guard would be one shaped like a comb as depicted in Figure 4.12.

The walls of each little vee are only visible from the corresponding shaded area and since these regions do not overlap, we need as many guards as we have vees. If there are *m* of these vees, then the total number of walls is $n = 3m$, so in this case the museum needs to hire $\frac{n}{3}$ guards.

The surprising result, revealed by a little network theory, is that the *m*-comb is as bad as it can get, meaning that in any musuem, the number of guards required is never any more than one third the number of walls. A simple proof of this was devised by Steve Fisk using network colouring. It relies on an idea frequently employed in mathematics to demostrate fundamental properties of complex shapes, that of *triangulation*.

In particular, if you draw a quadrilateral of any shape you will be able to split it into two triangles by drawing a diagonal between one pair of opposite corners. A five-sided shape can be similarly triangulated with two such diagonals, a six-sided figure will require three diagonals and in general any polygon, that is to say the plan of any museum, can be triangulated in this way by drawing $n - 3$ *non-crossing* diagonals between corners of the *n* walls. A formal proof of this fact relies on what is known as an *induction* argument, where we

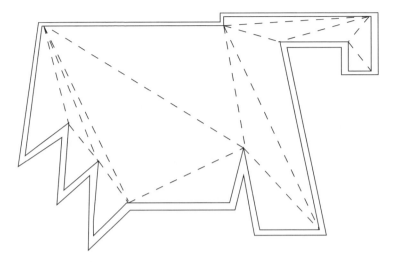

Figure 4.13 Triangulation of a gallery with 18 walls

build from one case to the next. It is slightly trickier than you might first expect as it takes some thought to show that it is always possible to find at least one diagonal that splits the polygon into two smaller ones with fewer sides. But it can be done.*

Taking this fact for granted, we can complete the proof with a piece of network magic. We think of the triangulated plan as a plane network in the obvious way. Figure 4.13 gives a representative example to focus on, where a museum with 18 walls is triangulated by 15 non-crossing diagonals. The nodes of the network are the corners of the gallery and the edges are the sides of the triangles used in the triangulation, some of which are the walls themselves. It turns out that this network is always 3-colourable. The reason this is true stems from the obvious fact that the chromatic number of a triangle is three. We may colour one triangle and then colour the remaining node of an adjacent triangle with a different colour from the two used to colour the ends of the common side of the triangle pair. Continuing in this way we may colour the entire network using just the three colours.

Now the number of walls, n, is the same as the total number of corners of the collection of triangles as we may match walls with corners as we move in a complete circuit around the perimeter of the gallery (because the triangulation does not permit new corners to arise as the diagonals used do not cross one another). Each corner now has a colour associated with it and so at least one of the three colours is used no more than one third of the time (by the Pigeonhole Principle). Stationing your guards at each of these corners, which are no more than $\frac{n}{3}$ in number, then gives full surveillance of the museum as every triangle has at least one guard, and so the space within every triangle is visible and therefore the entire gallery is covered.

2. Innocent questions of points and lines

We next feature a well-known result that amounts to just a single very simple observation about points and lines: given any finite collection of points in the plane, not all lying on the one line, it is possible to find a line that passes through exactly two of them.

A little experimentation with dots and lines will soon convince you of this fact, but how can we be sure? The proof must require some thought as the claim fails if we drop the proviso that the collection is finite: if we take *all* points in the plane then every line contains infinitely many of them!

Part of the reason the problem is interesting is the surprising variety of ideas that have been brought to bear to prove it and similar results along these lines. As you will be expecting, here we give a proof based on planarity. However, the dots and associated lines will not in general give us a planar network and no progress lies in that direction. Nonetheless, N. Steenrod showed that the problem was amenable to this kind of argument if it is transferred to another setting. This approach of introducing a 'magic mirror' is one that mathematicians are very fond of. Sometimes, for reasons that can remain mysterious, the solution to a problem becomes clear when it is transformed into a different environment.

The particular transformation used in this instance involves passing from the plane to a sphere. Up until this point we have always drawn our networks on a flat sheet of paper, that is, in a plane.

The same net of connections can be represented on other surfaces and a network that is not planar on one surface can sometimes be drawn on another without the edges crossing. In particular, surfaces with one or more holes in them, such as a doughnut, allow things to happen that are impossible in the plane. For example, the two basic non-planar networks, K_5 and $K_{3,3}$, can both be drawn on the surface of a doughnut (known as a *torus*) in a plane fashion. This may surprise you but, if you have such a shape to hand, or even if you draw one, it is not hard to see how to use the hole to avoid the crossing that is unavoidable when we keep everything flat. The other side of the coin is that you need more colours to colour maps drawn on a torus. Indeed on the surface of a doughnut it is possible to draw up to seven regions with each pair of them having a common border.[1]

However, the plane networks that can be depicted on a sphere are just the same as the ones that work in the plane. It is clear that any plane network can be transferred to a sphere and planarity is retained—indeed if you make the sphere big enough compared with the picture of your network, the surface appears almost flat and so the picture is hardly distinguishable from the plane representation. We can also go in the other direction: suppose you draw a network on a sphere with no edge crossings. This network can then be smeared into a very small part of the surface of the sphere without disrupting the connections in your network. We can imagine taking a small circle within one face of the network and letting the circle expand, pushing the edges and nodes of the network along with it as it grows. Eventually the network will lie within one hemisphere and we can then contract the network futher until it lies within a small circle on the sphere's surface. Eventually the plane network will occupy a small section of the sphere that is almost flat and this configuration can then be projected onto a flat plane with the planarity of the network still intact. The sphere gives us no more 'planarity power' than does the the plane. We cannot use the nature of the sphere somehow to avoid the otherwise unavoidable edge crossing in a network such as K_5.

[1] See for example the webpage <http://enderton.org/eric/torus/omdex.html>.

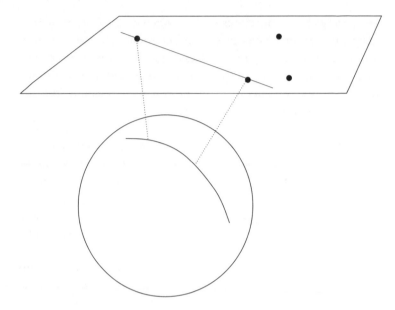

Figure 4.14 Projecting lines in a plane onto a sphere

All the same, transferring the Sylvester–Gallai problem, as it is known, from the plane to the sphere helps solve it, but not in a fashion you might naturally expect. We transfer the points and lines in the plane by projecting them on to a nearby sphere as suggested by Figure 4.14.

We imagine the plane suspended above a sphere and identify each point in the plane with the *diameter* of the sphere whose extension passes through the point in the plane on which we are focusing. This diameter is determined by the line through the point and the centre of the sphere and intersects the sphere's surface at two points exactly opposite one another on the globe, known as *antipodes*. In this way we may regard the point in the plane as being identified with this *pair* of antipodal points on the sphere or, if we prefer, we can think in terms of the diameter bounded by the pair of antipodes. In any case, imagine the point in the plane moving along a straight line. As the point traces out the entire line, in both directions, the corresponding diameter of the sphere performs a 180° turn, tracing out a *great circle* on the surface of the sphere, that is a circle with its centre at the centre of the globe. If you like to think in terms of globes of the Earth, lines of longitude are exactly the great circles that go through the pair of antipodes represented by the north and south poles. Circles of latitude, however, with the exception of the equator, are not great circles for although their centres lie on the polar axis, they do not lie at the centre of the planet.

And so we can recast our points and lines in the plane as diameters and great circles of the globe: a collection of points in the plane, not all on one line, give rise to a set D of diameters of the globe, not all being the diameters of the same great circle.

Now comes the trick. We recast the problem as it appears on the sphere in another way. With each diameter from D we associate the great circle whose plane is at right angles to it—for example, if the diameter happened to be the north–south axis, then the great circle in question is the equator. There is nothing special about this axis, however, for any diameter there is such a great circle. Let us write G for this collection of great circles that arise from the members of D. If the diameters of D *were* all common to one great circle, this would correspond to the set G of great circles that arise from these diameters all having a common diameter as well, that diameter being the one at right angles to the common great circle: that is to say all the circles of G would share a common axis through a pair of antipodal points. However, since this is not the case, the associated great circles of G do *not* all share a single common diameter.

The required conclusion of a line containing exactly two of the original points corresponds to a great circle that has exactly two members of D as diameter, which in turn corresponds to a pair of great circles from G whose common diameter is not shared with any other member of G. It is this version of the conclusion that we now demonstrate.

This conclusion drops out by treating G as a planar network on the sphere with the nodes being the points where great circles meet and the edges begins the arcs of the great circles between these common nodes. This network is simple—multiple edges could only arise if all the great circles went through a common pair of antipodal points, which is precluded as not all of the original set of points lay on the same line.

In this planar network, as a great circle passes through a node, it contributes *two* to its degree and so all nodes are even and indeed, since a node represents a point where great circles meet, its degree is at least four. Since the network is planar, *it must have a node of degree no more than five*—in this context that degree must be no more than four, and so by the previous comment, *there must be a node of degree exactly four*. That is to say, there is a point on the sphere where exactly two, but not more than two, of the great circles meet, which is the conclusion that we seek.

Remarkable and short as this argument is, it is natural to wonder why the proof has come in so roundabout a way. Surely there is a simple and direct way of demonstrating this fact?

The answer is 'yes' and here is a proof that avoids introducing spheres or networks and grapples directly with the points and the lines involved. Let P denote the set of given points and L the set of all lines that pass through two or more points of P. For each line l in L, consider the points p in P that do *not* lie on l: there is always at least one such point p for any l as we are given that not all the points lie on the one line. Of all these pairs of lines and points, choose one pair, l_0 and p_0, such that the distance from the point to the line is as small as possible. (This is possible because we have a finite collection, so that this minimum is attained.) We shall show that the line l_0 contains exactly two of the points of P.

All the lines in L, including l_0, contain at least two points of P but suppose, contrary to what we want to show, that l_0 had three points of P. Let q be the closest point on l_0 to the point p_0 (see Figure 4.15). Then at least two of the three points lie on l_0 on the

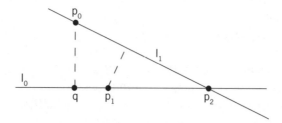

Figure 4.15 Proof of the Sylvester–Gallai theorem

same side of q, with the order of these points on the line being q, p_1, p_2 say (although p_1 might actually be equal to q). This however leads to trouble, for look at the line l_1 that goes through the points p_0 and p_2. Since l_1 has two points of P, the line l_1 is one of the lines in our collection L. On the other hand, the distance of the point p_1 to l_1 is less than that of p_0 to q, which contradicts the way we chose the point p_0 and the line l_0 in the first place. This contradiction is the inevitable consequence of supposing that l_0 contained more than two points from P, and so it is this that must be wrong. Hence our claim that there is a line that contains just two and no more of the given points is valid and so the Sylvester–Gallai Theorem, as this fact is known, is true.

This straightforward proof may strike you as more natural and easier to remember than our first argument. All the same, upon reading it, a mathematician might feel a little uneasy and suspect that there remains more to be said on the question. The proof is short and clear but it does make crucial uses of the idea of distance between points. There is nothing wrong with this, but it is an aspect of the question that we might have thought would not need to feature heavily in a proof of this result. This contrasts with our first proof that did not rely on any specific comparison of distances between points.

In finite geometries at their most abstract, there is a notion of points and lines incident with each other, but no notion of distance, angles, or even of order. However, in these very general settings, the Sylvester–Gallai Theorem simply does not hold so that it seems that any proof requires some additional structure to work with, such as a meaningful order for the points on a line.

Members of the general public are often bemused as to why mathematicians fret about this kind of thing. However, it comes about because mathematicians are not satisfied with any old proof as their mission is to understand all that surrounds the question as thoroughly as possible. Although a vague and elusive goal, this is a very important part of the motivation for real research. It may sound nonsensical for a mathematician to 'disapprove' of a proof. What is meant by such a criticism is not that the proof is invalid but that the line of argument is not the best one to take. Another approach might clear up the question thoroughly and lead to more enlightening ideas. A 'bad' proof on the other hand might be a dead end that has the effect of inhibiting further progress rather than fostering it. On the one hand, favourite techniques can be a matter of taste but on the other, some approaches can prove more fruitful than others and, in time, these subjective questions can, to a large degree, become settled.

3. Brouwer's fixed point theorem

All of us like to believe from time to time that life has some fixed points. We would have it that there are some good things that never change no matter what upheavel may strike our own world. Mathematics itself is one of these everlasting wonders and what is more, it offers some support to this optimistic outlook through the particular topic of *fixed point theorems* that gain application in a variety of subjects, especially in economics.

The simplest fixed point theorem applies to a line segment that is mapped into itself in a continuous way. We take our segment to be of unit length so that the left-hand endpoint is numbered 0 while the right-hand endpoint carries the number 1. Intermediate points are labelled by x, where x is the distance as measured from 0. Let us write $f(x)$ for the value of the point to where x is moved to on the line segment under some *continuous* transformation, where by continuous we mean that the interval may be stretched in places, compressed in others, and even folded onto itself (so that several points are mapped on top of one another), but the domain is not torn. Another way of viewing continuity is that neighbouring points are mapped to neighbouring points—it is not possible to find points arbitrarily close to one another that have images separated by some fixed value.

We intend to convince ourselves that at least one point p is left untouched by all this moving about so that $f(p) = p$. If neither of the endpoints remain fixed then we have that $0 < f(0)$ (0 is less than its image) and $f(1) < 1$. Imagine what we would see if we were to draw the graph of $y = f(x)$ on ordinary cartesian (xy) axes. Since we are insisting that our *function* f is continuous, its graph would be some kind of continuous (although perhaps extremely irregular) curve. We can say little more about it in general *except* that it will begin *above* the diagonal line $y = x$, because $0 < f(0)$, and finish *below* the same line, as $f(1) < 1$. Therefore, somewhere in between, it must meet the line $y = x$ at some point, (p,p) say, and so $f(p) = p$ and we have found our fixed point.

To be honest, we have not *found* our fixed point at all, but rather have deduced that there must be one somewhere between 0 and 1 because the graph, being that of a continuous function, cannot jump from one side of the diagonal line to the other. Indeed it may cross back and forth, cutting the line any number of times and even coincide with it for a time, which would give us infinitely many fixed points in the interval. We can be sure, however, that there is some point p in that interval where, like baby bear's porridge, the balance between x and $f(x)$ has to be just right, and so p and $f(p)$ are precisely the same value. To use a word of which mathematicians are fond, the position of p is *invariant* under this transformation.

Does this apply to a disc? Imagine a circle and rotate it about its centre through some particular angle, let us say, in order to be definite, anti-clockwise through a right angle. Every single point on the face of this disc is moved under this transformation *except* the centre of the circle, which remains where it was. Here then is a simple example of a continous transformation of a disc with only one fixed point. In 1910 L. E. J. Brouwer (1881–1966) published a paper, however, in which he proved that a fixed point in this and similar situations was inevitable. Indeed his theorem applies not only to discs but to any region that could be continously transformed into a disc and indeed not only in two dimensions. Brouwer's theorem proves that if a sphere, or a region that could be deformed in a continuous way into a sphere, in any number of dimensions, is mapped back into itself in a continuous manner, there is then a fixed point.

A graphic illustration for this theorem applies to two identical sheets of paper, one laid over the other. If the top sheet is then crumpled up any way you wish (without tearing) and placed on top of the other then there is at least one point of the crumpled sheet that lies exactly on top of its original position before the crumpling took place. The first sheet, although a rectangle perhaps, corresponds to the original disc, while the second sheet represents the action of the continuous function: a point p on the second sheet whose original co-ordinates were (x, y), say, now lies over a new point p' on the first sheet with new co-ordinates (x', y'). However, Brouwer guarantees that for at least one point p, $x = x'$ and $y = y'$.

However, you don't know where this fixed point p is, and Brouwer's proof does not show you how to locate it. In his later life, Brouwer himself found this profoundly unsatisfactory and personally rejected this kind of existence argument. He preferred to regard a mathematical object as only truly existing if it could be constructed. For example it might be possible to prove that there exists a 10×10 Graeco-Latin square (see Chapter 2) without specifically constructing it. This might be done by showing that the non-existence of such a square leads to a contradiction. Brouwer's later mathematical philosophy would not deny that a contradiction had been correctly deduced but he would not accept that we had a proof that such a square exists until one had been written down. More precisely, the constructionist approach would insist that we devise an algorithm for producing the square that would thereby guarantee that we *could* produce the square after some calculation that is bound to be of finite duration. Brouwer would not insist that you always need to go through the calculation in question but if you did, more's the better.

Most mathematicians do not share Brouwer's view but have some respect for it all the same. A contradiction argument would be accepted as a proper proof of the existence of an object even if it gave no clue as to how to find such an object. However, the proof should then act as a spur to the mathematical community to go out and find what they know must be there.

In the case of an example such as Graeco-Latin squares, there would be less controversy, as in principle, all possibilities could be enumerated and checked so there is an algorithm at hand for settling the question, even if it is too difficult to implement in practice. However, existence arguments involving infinite collections can leave one wondering and it is not hard to give a real example.

A number that is a simple fraction, such as $\frac{5}{8} = 0.625$ is called *rational*. Rational numbers are characterized by having decimal expansions that either terminate or fall into a recurring patterns such as $\frac{5}{12}$ = 0.416666 .... Numbers that lack such recurring expansions, such as $\sqrt{2}$ and π, are called *irrational*. It is easy to show that *there exist* two irrational numbers, a and b, such that a^b is rational. To see this consider the number $c = \sqrt{2}^{\sqrt{2}}$. Either c is rational or it is not. If it is, we have already found an example by putting $a = b = \sqrt{2}$. If not, put $a = c$ and $b = \sqrt{2}$. Then $a^b = (\sqrt{2}^{\sqrt{2}})^{\sqrt{2}} = \sqrt{2}^2 = 2$, which is rational. Since one of the two cases must apply, the existence of the required numbers is proved.

This is just the kind of proof Brouwer would not have had any truck with. The technicalities of the argument are not important here. Rather it is the fact that it gives two alternatives yet provides no clue as to which one applies. It merely observes that if one does not work then, as a consequence, the other must. (Fortunately this particular problem has been settled: it is known that $\sqrt{2}^{\sqrt{2}}$ is irrational, but the proof is very difficult.) However, Brouwer is right in pointing out that we are not really that much the wiser for such a proof as it leaves us with no way of testing, not even in principle, which of the two alternatives applies. The proof is of real value all the same—many students of mathematics might mistakenly think that it is impossible for an irrational power of an irrational to be rational, for it just doesn't sound right. This little argument at least tells you not to waste your time trying to prove that, even if it does not settle the question to the extent we would like to see.

Paradoxes

Most of us would still not have too much sympathy with Brouwer's pedantic view as we all like to believe that any statement is either true or false with no third option. Accepting this allows us to make use of contradiction arguments to settle questions for if we prove that a statement leads to contradiction, then its opposite must be true and the question is decided. However, it is just not possible to adopt this attractive commonsense viewpoint in a totally unqualified manner and remain consistent.

The dilemma as to what constitutes a real proof can be traced right back to that disturbing utterance of Epimenides the Cretan in 600 BC that 'All Cretans are liars.' (See Chapter 1.) Self-reference seems to lie at the heart of all inconsistency and contradiction, something that has plagued both mathematicians and philosophers through the ages and has never been resolved—the existence and eventual non-existence of our own consciousness seems to smack of the same dilemma whenever we ask ourselves the unanswerable question, 'Why am I me, and not someone else?'

The mathematical difficulty arises when we naively assume that any statement is either true or false. We may not know which is the correct alternative, conceivably it is impossible ever to find out, but surely one or the other applies. Epimenides alerted us to the difficulty of this position with his statement about Cretans. To put the paradox more baldly, consider the sentence, 'This statement is false.' If we assume that it is true, then it is false (for that is what it says), and if we assume it is false, then we infer that it must be true. We conclude that we simply cannot assign a *truth value*, to this 'statement'.

This is an annoying example. It is self-referential and indeed refers to itself as an existing statement before it is even completed. We could argue that this is implicitly nonsensical and so this kind of thing should be prohibited. Having said that, it seems we can return to our commonsense position that any (sensible) statement is either true or false, and there is no third alternative.

All the same, a spanner has been thrown in the works, for how are we to define and recognize these kinds of nonsensical statements? Whatever definition we come up with may conceivably leave the door open to other assertions that also cannot be given a truth value. How can we be sure we have banished all troublemakers?

We could therefore be driven to issuing a blanket ban on any statement that causes trouble in this way. That would be consistent but would leave us very much in the dark, for how could we know in advance whether a statement led to contradiction whichever truth value is assigned to it?

Nonetheless, something along these lines has to be done to rescue logic and mathematics. It is, however, a disappointment to be forced into quite technical considerations on a matter where all seems clear except for unimportant exceptions that are deliberately designed to be a nuisance. For this we are forced to relinquish the absolute freedom of expression we thought we had by rights. A similar dilemma arose with a vengeance in the nineteenth century when *Set Theory* came to the fore.

The idea that every collection of mathematical objects could be regarded as a *set* was a notion that freed the subject, allowing it a single arena where all mathematics could be carried out. This 'paradise' as David Hilbert described it soon started to beget contradiction nevertheless, along the lines that Epimenides had warned us about all those years ago. Although not the first paradox of set theory, that of Bertrand Russell is perhaps the most famous and shows that if we believe we can define sets in any fashion whatsoever, we land in trouble.

Russell's paradox concerns the set S of all sets that are not members of themselves. It is possible for a set to be a member of itself— the quaint example that Russell himself uses is the set T of all things that are *not* teaspoons. Whatever T is, it is evidently not a teaspoon, so that T is itself a member of T. We write this symbolically as $T \in T$, the funny $\in$ sign being shorthand for 'is a member of' or 'belongs to'. Returning to Russell's set S, the embarassing question to ask is:

$$\text{Is } S \in S?$$

However you answer yields a contradiction: if S is a member of S, then S must meet the qualification to be in S, which is that S is *not* a member of itself, and so that can't be right. Therefore the other alternative must apply: S is not in S; but since S must fail to meet the entry requirement for membership of S, it follows that S is a member of itself after all, and so $S \in S$ must be true.

It simply does not work and so in order to have a consistent theory of sets we need to introduce restrictions on how a set can be defined. This is what Russell and others have done. There really was no choice other than to restrict the theory in some way or other, but again it is uncomfortable for there is more than one way to construct a reasonable theory. Which should we choose and how can we yet be sure our theory is consistent? Some mathematicians make a living from sorting these things out, while many still don't care or at least take the attitude that they won't worry about foundations until they somehow cause trouble for them personally.

Sperner's Lemma

Leaving philosophical musings aside, the purpose of introducing Brouwer's theorem is that it has a remarkable proof based on a remarkable fact known as *Sperner's Lemma*. A lemma is a mathematical theorem, sometimes of a technical nature, which allows you to go on to deduce more interesting things. However, to a mathematician, the proof of a key lemma often represents the heart of a topic in that it identifies the driving imperative behind a whole body of interesting theory. Sperner's Lemma is a true lemma and, most interesting for us, the lemma itself is a surprising by-product of a problem of network colouring.

John Fraleigh, in his classic undergraduate text on abstract algebra, tells his students never to underestimate a theorem that counts something. This mathematical maxim needs a little explanation as the idea of a theorem that 'counts' goes a little wider than discovering a formula for finding the number of certain mathematical objects, although that would certainly be included within the scope of counting theorems.

An argument that leads to a conclusion such as the number of objects of a certain kind is odd, or is a multiple of three, or is a prime is often very powerful as it can place great constraint on what is possible. A simple example arises in the form of so-called *parity arguments* in which two very similar-looking objects are shown to be fundamentally different by identifying one feature that is 'even' in one object but 'odd' in the other, and so they cannot be the same.

This can all be made clear by simple examples. A classic instance comes from one of the celebrated puzzle books of Martin Gardiner: the problem of the mutilated chessboard. Suppose we have a chessboard and also some dominoes each of which is just the right size to cover two squares on the board. We can cover the board with these dominoes very easily, laying down four in each row for instance. If we now mutilate the board by cutting out two diagonally opposite corner squares, can the remaining board be covered by the dominoes? (Without, of course, any protruding over the edge.)

The answer is 'no' and this can be seen by focusing on the *colours* of the remaining squares. The two squares removed are necessarily of the *same* colour: let us suppose they are both white. The remaining board then has two more black squares than white. However you lay a domino on the board, it will cover two adjacent squares, one of which will be white, the other black. It follows that as we cover the board with dominos, at any stage we will have covered equal numbers of black and white squares and so the covering by dominos and the entire mutilated board can never be made to match.

Another similar example of a parity argument concerns a teacher's class in which 35 children are seated in a 7×5 rectangle. She wants to rearrange the seating so that every child moves to a new desk but, to make things simple, she orders everyone to move just one space, either forwards, backwards, or sideways. She then tells the children to sort it out among themselves.

The teacher has set her poor class an impossible game of musical chairs. How can we tell? Once again imagine the array as a mini chessboard, with alternate spaces coloured black and white. The trouble is, there are an odd number of squares, so the number of black and white squares will differ by one—if we colour the first square white we will end up with 18 white and 17 black. Now when places are changed in the manner prescribed, a child in a white square moves into a black one (and vice versa). However, since there is one more white square than black, one child will always be left with nowhere to sit. Very frustrating for the poor kids but they are not being silly—their teacher has inadvertently set them a problem with no solution.

Returning to the matter in hand, the key to our verification of the Brouwer Theorem is analysis of the following situation, somewhat akin to our problem of guarding the gallery. Suppose that we have a big triangle with vertices V_1, V_2, and V_3 and it is triangulated as in our earlier problem. A typical outcome could be as pictured in Figure 4.16.

We now colour the vertices of the network with three colours 1, 2, and 3 but not according to the usual rule that adjacent colours need be different. There are rather different constraints, these being, V_1, V_2, and V_3 are coloured 1, 2, and 3 respectively, the vertices on the side from V_1 to V_2 are only coloured 1 and 2, and similarly the side from V_2 to V_3 only carries the colours 2 and 3, and likewise the 3–1 side is only coloured using 1 and 3. The interior vertices

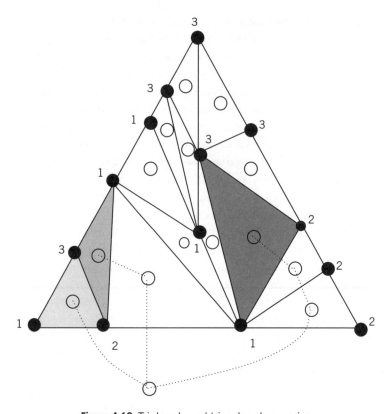

Figure 4.16 Triply coloured triangles always arise

of the triangle however are coloured using 1, 2, and 3 with no restriction whatever. Sperner's Lemma then concludes that a triangle carrying all three colours must emerge within this triangulation.

As you see, this is a true 'lemma' as the set-up looks quite artificial and the conclusion seems to be of little or no interest. However, that is an illusion as this little result is very powerful. And we prove it in a very sneaky way—we show the the number of tri-coloured triangles must be odd. (And so there must be at least one of them!)

The idea is to draw a kind of map network but subject to different rules. As with the map network we have one node for each triangle and another for the outside. However, two nodes are joined by an edge only if their separating edge is labelled 1–2.

Let us now look at the degree of each node. The outside node is odd as there is one edge from the outside node through each of the edges on the base of the triangle labelled by both 1 and 2, and since there is an odd number of alternations between 1 and 2 on the base (as we begin with a 1 and end with a 2), then the outside node must be odd. A node will be isolated, and so of degree 0, if its triangle lacks either of the colours 1 and 2; if a node lies within a triangle with colours 1 and 2 but no 3, it will be of degree 2; and finally any triply coloured triangle will have degree 1. We now finish the proof by invoking the Hand-Shaking Lemma—the number of nodes of odd degree is even, and since the outside node is odd, there must be an odd number of other nodes of degree 1, that is to say there is an odd number of triply coloured triangles. Therefore Sperner's Lemma is established.

All this can be seen in action in the example of Figure 4.16. In this instance the number of triply coloured triangles is three and they are shaded.

Using Sperner's Lemma, it is surprising but true that it is quite easy to deduce Brouwer's Theorem for the disc and the details are recorded for interested readers in the final chapter, although a little additional mathematical knowledge is required.*

The first observation is that we can work with an equilateral triangle instead of a disc because one may be continuously deformed into the other, and so that a mapping of the disc that was fixed-point free could be used to produce a fixed-point free mapping on the triangle. To show that no such mapping of the triangle can arise, the argument goes by way of ever finer triangulations of the triangle itself and a timely application of Sperner's Lemma to contradict the assumption that no point remains unmoved.

5

How to Traverse a Network

L et us now return to Euler's resolution of the question of finding traversing walks in networks. Our explanation is thoroughly modern and is not expressed in the fashion that Euler would have used. Although he is regarded as an excellent expositor of his own work and is renowned for producing mathematics that was free of error, he suffered greatly from the fact that network theory was a new and unrecognized field. It would have been strange in the extreme to devise the kind of language that we have introduced here, incorporating as it does a large number of interrelated ideas, simply in order to deal with one or two problem types. Lacking this language and notation was however a great handicap and Euler struggled to convey his new ideas, despite the fact that, to the modern mathematical outlook, they are not especially difficult to come to terms with and put into practice.

The Euler–Fleury Method

Recall now that an *Euler path* in a network is one that traverses each edge exactly once and let us call such a path an *Euler circuit* if it is both an Euler path and a circuit. For example, the bow-tie of Figure 3.7 has an Euler circuit that can be described as: $A \to B \to C \to D \to E \to C \to A$.

To traverse a network N, it must be *connected*, that is to say, must consist of just one component, so let us take that for granted. Euler laid down the law on when you can traverse a connected network. The full story is:

1. N can be traversed by an Euler circuit if all nodes are of even degree, and not otherwise;
2. N can be traversed by an Euler path, but not by an Euler circuit, if it has two odd nodes and not otherwise. Moreover, any Euler path must begin at one odd node and terminate at the other.

We have already explained why these criteria must be met. Each time an Euler path passes through a node it uses up a pair of edges that we are forbidden to use again. It follows that all the nodes must be even except perhaps the first and last. In a circuit, no node is intrinsically first or last (we must begin somewhere, but all nodes are equally good) so that no odd nodes are possible at all. Since we now know, as a corollary of the Hand-Shaking Lemma, that it is impossible to have a network with exactly one odd node, this is a case that never arises.

Deciding whether or not a network has an Euler circuit is a global question concerning the array as a whole. It does however have a local solution in that the question can be decided through inspection of local features of the network, namely the degrees of the nodes. In contrast, determining whether or not your network has a Hamilton cycle cannot always be decided through some series of local inspections.

Moreoever, and this contrasts with many other problems in network theory, the task of *finding* a solution to the Euler circuit problem is relatively easy, although not as easy as it might be, for a completely naive approach can let you down. Suppose for instance that the network has exactly two odd vertices, u and v. We might begin at u and cross any available edge, without giving it a thought, and keep going, hoping for the best. If you are careless, however, you can get stuck and find yourself sitting on a node with no way out, even though you have not yet traversed all the edges.

For example, the network of Figure 5.1 has exactly two odd vertices, numbered 1 and 8, and so we should be able to traverse all the edges of the network, beginning at 1 and finishing at 8. However if we begin our walk with $1 \to 2 \to 3 \to 6 \to 7 \to \ldots$, we have landed

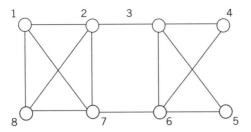

Figure 5.1 A traversable network

ourselves in trouble. If we imagine burning our bridges as we pass over each one, we see that upon our arrival at 7, the remaining network has split into two pieces and we have managed to strand ourselves on the left-hand side, with no prospect of traversing the edges still unwalked in what remains within the right-hand component. This is the only difficulty that can arise, however, and it is readily avoided. We do not have to be very clever when we construct our walk—we don't have to think two steps ahead—we only need to avoid taking a step that splits what is left of the network into two pieces. We can indeed give an automatic procedure that will work and avoids the necessity of guessing and hoping. This solution to the problem is due to Fleury.

To traverse a network with no more than two odd nodes, begin at any vertex you wish if there are no odd nodes, and at either of the two odd nodes otherwise. You may now walk the edges of the network making sure that:

1. You draw a picture of the network and erase as you go any edge that you have used and any node that has had all of its edges traversed;
2. At each step use a *bridge* (sometimes called an *isthmus*), that is to say an edge connecting two otherwise disconnected components of the remaining network, only if there is no choice.

You should have no trouble traversing the above network now, starting at 1 and ending at 8. Note that the failed walk above violated rule 2 when the choice was made to travel along the edge 6 → 7 as this crosses an isthmus of the remaining network.

A list of instructions like this one is known as an *algorithm*. It gives a mechanical procedure, one that in principle could be programmed on a computer, by which the problem can be solved. Having a simply described algorithm that solves a given problem is not always the end of the matter. If the process that the algorithm demands takes an impossibly long time to carry out, it may be of no practical use. For this reason no end of effort still goes into finding faster and faster algorithms to deal with problems that have already been 'solved'. However, the above example for finding Euler circuits is a good algorithm in that it can be implemented on even large networks and the procedure is genuinely feasible.

An explanation showing that the Fleury method must find a traversing walk whenever one exists is in the final chapter.* (After all, I have provided no proof that the Euler condition of 'no odd nodes' is sufficient to guarantee the existence of an Euler circuit, nor that the preceding algorithm always works.) This kind of problem is popular in riddle books but is usually phrased in terms of drawing the bridges in rather than rubbing them out. The question often posed is: Can you draw this picture without taking your pencil off the page and without going over any line twice? The standard pair

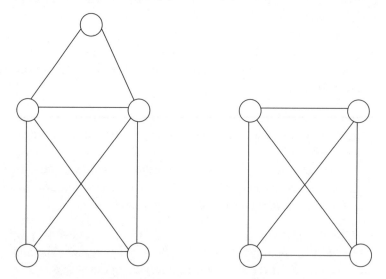

Figure 5.2 Traversing the open and the closed envelope

of examples that can and cannot be done respectively are the open and closed envelopes, shown in Figure 5.2. Provided you begin and end at the bottom, the first figure can be drawn within the rules, but the closed envelope, with its four odd vertices, is impossible. And as mentioned in Chapter 2, there are variants, such as traversing through all the doors of a house.

To be able to solve a network problem through the application of a few simple rules is not something that can be taken for granted, but there are other interesting problems that succumb to simple techniques, as we shall see.

The Chinese Postman Problem

A very practical problem that all of us who have ever had to deliver goods or services to many dwellings will have met is how to plan the route so as to avoid unnecessary traipsing about. It is most easily understood in terms of a postman who has to perform a mail run where he leaves and eventually returns to his post office base having delivered to every street in a section of town for which he is responsible. The problem was first considered as early as 1917 by H. E. Dudeney but it now always goes by the name of the *Chinese Postman Problem* because the complete solution was devised and explained by the Chinese mathematician Mei-ko Kwan in a short paper in 1962.

It comes down to a problem about Euler paths. Indeed the postman's delivery area can be imagined as a network where we place a node on each corner and join corners by edges if a street runs between them. From this we can see that the postman already has a solution to hand if he is lucky enough to have a street network where every street corner represents an even node. All he needs to do is find an Euler circuit that begins and ends at his base through using the method outlined above. This will lead to him traversing each of his streets exactly once. This is as good as a solution can be as it involves no backtracking or repetition of edges in his walk.

However, what if there are odd nodes in the postman's world? He still needs to begin and end at his post office and now there is no Euler circuit so he will have to tolerate a walk that is not an ideal circuit free of any repetition. How can he find an optimal route?

Before plunging further in to this question, however, it is best to draw your attention to a guiding principle that is often used by mathematicians. When a problem is met that looks similar to but harder than another problem that you can deal with, a natural approach is to try to perturb the new problem so that it resembles the doable one, and then somehow manipulate the solution of the simpler problem into an answer to your new question. This rather vague advice would not be worth calling upon if were not the case that this kind of thinking is used so very often, and the Chinese Postman Problem is a case in point where it does indeed apply.

Think back to the problem of how to find an Euler path for a connected network N with exactly two odd nodes, u and v. Given that we can find an Euler circuit in the case of a network with only even nodes, we consider how we can associate a network of this type with N. If you were to draw your N on a sheet of paper you may well have a natural impulse to 'cheat' by drawing an extra edge between the offending nodes u and v to give yourself a network N' free from troublesome odd nodes. Rather than retreating in shame from this temptation, it is better to pursue this mischievous line of thought. Construct an Euler circuit for N' that begins with the new edge uv. The remaining part of the Euler circuit $v \to \cdots \to u$ then represents the required Euler path in the original network N. In this way we have simplified the case of a network with a pair of odd nodes to that of the standard case where all nodes are even.

Our postman can apply this thinking for the delivery network of Figure 5.3. This is a particularly simple problem as we assume all the streets have the same length. The post depot is at P but our postman does have two odd nodes, u and v, to contend with. The idea is to join the nodes u and v but, owing to the nature of the situation, this should not be done directly as there is no street that leads directly from u to v. Instead we adjoin two new edges, shown as dashed arcs

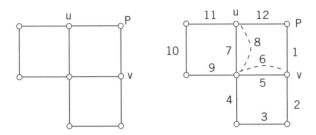

Figure 5.3 Simple Chinese Postman problem

in the picture, which will be used to retrace the streets in question. In this augmented network, the nodes u and v have become even and the intermediate node that has been affected remains even also as an even number of edges (two) has been adjoined to it: one for going in and the other for leaving. We can now find an Euler circuit in the new network, beginning and ending wherever we wish, so that the postman can begin and end at his depot P, and one such Euler circuit is indicated in the diagram on the right through the numbering of the edges. This must be an optimal solution to this problem: each odd node will require one of the adjoining streets incident with the node to be retraced and it cannot be done with a single edge because the nodes in question do not have a street that connects them.

This exercise is enough to illustrate some of the facets of the general solution but the question remains as to exactly how we would go about solving a problem of this kind faced with a really big network with lots of odd nodes scattered about. To see all the ideas in practice we need an example that is a bit more challenging (see Figure 5.4).

Let us insist that our postman walks all the streets shown, starting and ending at A. This time the streets have differing lengths, as shown by the edge labels, and he also has four odd nodes to cope with, A, B, C, and D. The first step is to group the odd nodes in pairs—this is something that can always be done as, by the Hand-Shaking Lemma, the number of odd nodes is always even. For example, we might try the pairings (A, B) and (C, D). If we adjoin an additional edge between the members of each of these pairs, which

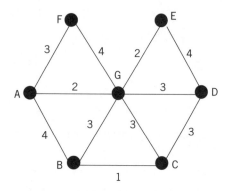

Figure 5.4 Postman facing four odd nodes

is possible in this instance as A is adjacent to B and C to D, then we have a network of even nodes. Any Euler circuit of this augmented network that begins and ends in A will be a possible route for our postman. Its length will be the sum of the lengths of all the streets on his route, plus the lengths AB and CD, which amounts to $4 + 3 = 7$ additional units.

Is this the best he can do? Not necessarily. After all, this method will give the postman a covering route for any pairing of the odd nodes and indeed we do have to check out every possible pairing. For each pairing, find the shortest path between the nodes of the pair and draw in the corresponding edges. The optimal routes are the ones where the additional mileage is as small as possible.

And so we continue to see if we can do better. Let us try the pairings (A, C) and (B, D) next. There are two equally short paths between A and C of length 5: the length of ABC is $4 + 1$ while AGC has total length $2 + 3$. The shortest path from B to D has length $1 + 3 = 4$. Putting in the reverse edges will lead to the postman walking an additional $5 + 4 = 9$ units above the total street lengths, which is two units farther than our first route.

Finally we calculate the effect of the third possible pairing: (A, D) and (B, C). We see that the increase above the basic street total in this case is given by $(2 + 3) + 1 = 6$, which is the best of them all. Here then is the solution to this particular Chinese Postman Problem: we draw in new edges corresponding to these paths: AG, GD, and BC and construct an Euler circuit for the augmented network. A solution is then given by

$$A \to B \to C \to D \to E \to G \to F \to A \to G \to$$
$$\to D \to G \to C \to B \to G \to A.$$

This represents the general method by which the Chinese Postman Problem is solved. However, the type of question involved goes somewhat beyond postal deliveries and snowplough routes. The labels on the edges can stand for any *weights* we wish, not necessarily physical distance. The most common alternative source of weighting in real-world problems is cost. Whether the weights represent costs or distances, mathematically the problem is identical, in that we minimize costs by minimizing the total weight of the route taken.

Although the previous example is indicative of a typical problem, we have cheated a little. One of the steps in the method tells you to find the shortest path between a given pair of odd nodes. You are entitled to ask, how do we do that? In principle it is easy. In principle we can simply list all paths between the two nodes and then choose one that is as short as possible. However, if the given network is very complicated, and real world networks are often extremely so, is that a feasible way to go about it?

The answer is 'no': the amount of computation involved in directly checking every conceivable path is prohibitive and a better way of solving this problem, the Shortest Path Problem, is called for. Fortunately, this is a problem that does have a short cut approach that always works. The method is known as Dijkstra's Algorithm and just how to work it will be shown in Chapter 7.

In conclusion, the Chinese Postman Problem does have a working solution, thanks in the main to Euler's original idea based on the Königsberg Bridges.

In a later chapter I shall show you a more recent and striking application of Euler circuits. In modern biology the problem arises of reconstructing an RNA chain, genetic material carried in the cells of a living organism, from the collection of fragments that result when the RNA chain breaks up in the presence of enzymes. The problem has certain idiosyncratic features that allow some inference as to the form of the original chain. After the decks have been cleared, however, the heart of the problem amounts to finding an Euler circuit in a certain directed graph the nodes of which are certain RNA fragments and whose arcs are indicative of other structural features. There is a one-to-one correspondence between the Eulerian circuits and the set of all possible RNA chains that lead to the observed collection of fragments. In other words, the network carries all the information that remains as to the original structure of the chain.

This application is a legacy of the ideas of Euler that he could never have anticipated. Seeing as it involves *directed* Euler circuits, however, its natural role in the story follows later.

6

One-Way Systems

All networks we have looked at hitherto have represented mutual relationships such as friendship or a two-way physical connection. However one-way systems are common. And not only in road and other transport networks: in physical and physiological systems involving valves, such as radios and the workings of the heart, we see traffic flowing in one direction only.

These directed networks are pictured in much the same way as two-way ones but now the edges carry arrows indicating the direction of the relationship between the two nodes. One-way networks like this are known in the trade as *digraphs*. This term is short for *directed graph* because a general network is often called a graph in the world of mathematics. Although we have stuck with the word 'network' throughout, we will, out of convenience, use the word digraph for these one-way networks.

One digraph that often arises is that of a *tournament*. The underlying network in this case is a complete network on some number of nodes and the directed edges, often known as *arcs* in the context of digraphs, can represent the result of matches in a round-robin tournament where each of a number of contestants all play one another once. An arrow from node *a* to node *b* indicates that it was player *a* who won the match between this pair. An example is to be seen in Figure 6.1 that could represent the results of a round-robin tournament of five players.

Ideas from undirected networks have obvious analogues for digraphs. For example, instead of simply speaking about the degree

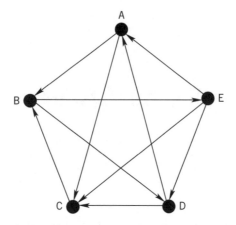

Figure 6.1 A tournament of five players

of a node we use the terms *out-degreee* and *in-degree*. For instance node E has out-degree 3 but in-degree 1, indicating that E won three of his four matches, only losing to player B. A node with only out-edges and no in-edges is called a *source* while a *sink* has only arrows going in, and no way out.

This particular tournament is *strongly connected* in that it is possible to find a *directed path*, that is one that respects the one-way system, from any node to any other. As a consequence, it can be proved that it is Hamiltonian in the sense that there is a *directed* cycle that passes through all the nodes of the digraph: $A \to C \to B \to E \to D \to A$; in particular, discovering this cycle shows that every node is accessible from every other. Not all tournaments are Hamiltonian: it is obviously impossible for a digraph to be so if it has a source or a sink, as there is no escape from a sink, while a source can never be reached from another node.

Since a digraph with a Hamiltonian circuit is obviously strongly connected, these two properties go hand in hand for tournaments. Although not every tournament is Hamiltonian, as we have already seen, it is the case that a tournament is always at least *semi-Hamiltonian*, meaning that there is a path that starts at one node, finishes at another (possibly different) node, and passes just once through all others on the way.*

The results of a tournament can be used to rank the performance of the players, but often not unambiguously. In our tournament, no one is undefeated but E is the top-ranked player, winning three out of four, while C won just the one match. The other three contestants are equally ranked with two wins a piece. In a round-robin tournament with n players, a complete ranking could only emerge if each of the possible number of wins: $0, 1, 2, \ldots, n - 1$ was achieved by the n players. There is a 'minimal criminal' characterization of tournaments which allow that. A tournament will yield a *perfect ranking* if and only if it does *not* contain a directed cycle of length three. In other words, a complete linear ranking will *not* be possible exactly when we witness a triangle of 'inconsistent' results as in our example where E beat C, C beat B, yet B defeated E.

One problem that naturally leads to a directed network is the construction of a one-way traffic system. Suppose that a town has a congested road network which is a free-for-all two-way system, that is to say every street carries traffic in both directions. Even if it involves longer journeys, traffic often flows better if the system is made one-way. This raises an obvious question: given a connected (two-way) network, can we orient each edge so that we obtain a one-way system?

The thing that can go wrong is that it may turn out that, however you go about it, the resulting directed network is not strongly connected—that is to say some places may be inaccessible if you start from the wrong part of town, and that would never do. And this can certainly happen. Imagine your town has a single bridge over the river. If that bridge took only one-way traffic, and if you happened to live on the wrong side of the bridge, you could never cross to the far side of the stream.

Indeed this is enough to let us see that we cannot make the system one-way if the underlying network has a bridge, which you will remember is an edge whose removal would leave two distinct components. However, this is the only aspect that ruins any possibility of a one-way system—as long as the road network is free of bridges (in the sense above) then it is feasible to orient the arcs in such a way that it is possible to travel from any place to any place else. It will be easier to explain how to do this however in the next chapter as the method involves spanning trees, which is the subject of Chapter 7.

Nets that remember where you have been

The analogue of an Euler circuit in a digraph is a directed circuit that covers every possible arc. Following closely the pattern of possibilities that arise in an undirected network, the principal result here is that there is a directed Euler circuit if every node is *balanced* in that there are as many arcs leading in to each node as there are leading out. This applies to one of my favourite types of directed networks called *de Bruijn graphs*. Part of their significance is that they are useful in the design of machine dial controls, a fact that requires a little explanation.

Figure 6.2 shows a rotating washing machine dial that allows 16 different settings. The setting in use corresponds to the four symbols on the top of the dial. Imagine that the sequence of four slots at the top of the picture is visible to the user and behind each slot there is a switch that can be either off (0) or on (1). As viewed in the diagram, the dial is in the off position with none of the switches active. As the dial is rotated, some of the switches become active, triggering a particular response from the machine. Since each switch has just two modes, off or on, the total number of different switch settings possible with the four switches is $2 \times 2 \times 2 \times 2 = 16$ in all.

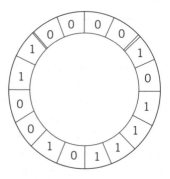

Figure 6.2 Dial with 16 settings

The dial will click through 16 positions as it rotates. What is remarkable here is that the cyclic sequence of 0's and 1's allows for each of the 16 possible settings to come up exactly once as we click the dial around one full turn. Specifically, if we click the above dial anticlockwise, the strings that appear in the top window come through in the following order:

0000 0001 0010 0101 1011 0111 1111 1110
1101 1010 0100 1001 0011 0110 1100 1000

That is to say this cyclic binary sequence, 0000101111010011, has the special property that starting from any point on the dial and reading, let us say clockwise, we find each of the possible 16 binary strings of length 4 coming up exactly once—very neat and efficient!

This kind of binary string is called a *de Bruijn sequence*. In general, a *de Bruijn sequence of order n* is a circular sequence on two symbols, often taken to be 0 and 1, of length 2^n in which every possible sequence of n consecutive digits appears. The above example of the washing machine dial features a de Bruijn sequence of order 4. For any circular arrangement of binary symbols of length 2^n, we can read 2^n binary strings of length n as we 'turn the dial', as it were. Since there are 2^n possible different strings of length n, if each appears once, then each must appear exactly once, as there is no room for repeats.

So we see that de Bruijn sequences are obviously good things to have, but is there any way of producing them? After all, they may not even exist—we can see with our own eyes that there is one of order 4, but what of order 5 and higher orders? Are they always there, ready for us to exploit, and if so, how can we generate them?

The answer lies in traversing all the edges of a de Bruijn graph, which we can now introduce. Figure 6.3 shows the de Bruijn graph of order 4.

The de Bruijn graph of order n has as its nodes the 2^{n-1} binary strings of length $n-1$. In Figure 6.3 we have $n = 4$ and so there are $2^3 = 8$ nodes labelled by the eight binary strings of length three. Each of these nodes has two out-edges, one labelled 0 and the other 1. What characterizes this graph and gives it a touch of magic is the rule that tells you where each arrow goes to: to locate the end

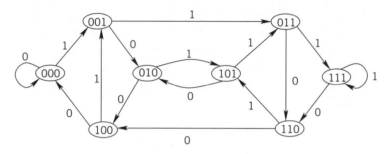

Figure 6.3 de Bruijn graph of order 4

node, take the label of your starting node, erase its first digit, and adjoin on the right the digit that labels the out-edge in question— that gives you the label of the node at the other end of the arrow. For example, apply this to the arrow labelled 0 that emanates from the node labelled 101: we erase the initial 1 and adjoin 0 on the right— the out-edge labelled 0 coming from 101 takes you to the node 010. Similarly the arrow labelled 1 from the same node terminates at 011. You will now see that this applies right throughout this network. We have described the workings of the rule for the case $n = 4$ but the same applies for any de Bruijn graph and is no more difficult to execute. For example, you may care to test it on the de Bruijn graph of order 5 that is drawn in Figure 6.4.

De Bruijn graphs are ideal for people who are easily lost. No matter where you happen to be in a de Bruijn graph you can always get to your home node, whatever it might be, by calling its name. From whatever node you find yourself at, just follow the path whose label is the name of the node you wish to find and it will take you there, as if by magic.

For example, in Figure 6.3 begin anywhere you wish and follow the path labelled 011 and you will finish at node 011. This remark-able property is built into the very design of the network: any path of length 1 goes to a node that ends in the digit labelling the path. When we extend to a path of length 2, the terminal node will record the label of the path in its final two digits, as the label of the first digit in the path will be preserved in the label of the end node, only shunted one place to the left. For illustration, take Figure 6.3 and

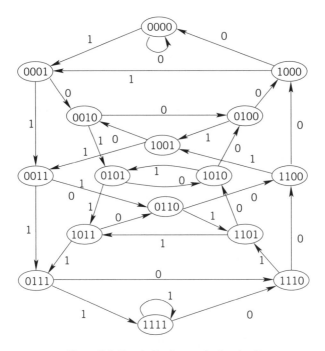

Figure 6.4 the de Bruijn graph of order 5

take any path with label 01; it will take you to one of the two nodes 001 or 101, no matter where you begin in the network. Similarly, if your path has length 3 (or more) the label of the final three edges will be recorded in the three digits of the terminal node. In the de Bruijn graph of Figure 6.3, that is the limit that the nodes can 'remember' but in general, the name of the terminal node in the de Bruijn graph of order n tells you the label of the path consisting of the final $n-1$ edges of that path. To further underline the point, look to the more complex order 5 de Bruijn graph (Figure 6.4). Any path ending 11 terminates at one of the four nodes 0011,0111,1011 or 1111, while you may check for yourself that wherever you begin, the path 0000 for instance leads you to the node atop the picture.

The secret to producing de Bruijn sequences for our washing machines is to take the label of any Euler circuit in the appropriate de Bruijn graph. For our machine with 16 settings, the graph of Figure 6.3 will do the job. Beginning at the appropriate node (explained in a moment), read in the circular label of our washing machine dial and you will find yourself traversing every directed arc of the de Bruijn graph exactly once and returning to your starting node. More importantly, we can generate de Bruijn sequences by writing down the label of any Euler circuit of the corresponding de Bruijn graph. A few experiments should be enough to convince you that all this works as it ought, but why should that be so?

Let us see what happens as regards Figure 6.3, which is quite representative of the general situation. The network has 16 arcs and each node has two coming in and two going out. It follows that the network will have an Euler circuit and the sequence of labels on the successive arcs produces a binary string of length 16. Moreover, that string will be a de Bruijn sequence and, conversely, any de Bruijn sequence corresponds to an Euler circuit of the digraph, as we now explain.

To see an example of this last point, let us take the de Bruijn sequence of our washing machine dial: 0000101111010011. This does provide us with an Euler circuit, provided that we begin at the right node. For example, if we start at the node 000 and read in this path we don't get an Euler circuit, as we can see right away, as we would begin by circling around the loop labelled 0 at this node four times before leaving it—definitely not we want to do. The clue as to where to start comes from the end of the sequence—specifically the last three digits, 011, tell you where you will finish so that node 011 will be the start and end of your Euler circuit. If you begin your walk from this node, you will find yourself taken by the string on an Euler circuit of the entire network.

To see this, we note that by design the walk determined by a given de Bruijn sequence will finish where it began after traversing 16 edges. We need only convince ourselves therefore that no arc is covered twice for, if that is the case, it follows that all 16 arcs must be traversed exactly once by the 16 arcs in our walk. To this end, then, take any arc e that is traversed in our walk. Suppose this arc exits from node abc, say, and is labelled by d, where each of the letters $a, b, c,$ and d represent either 0 or 1. By the nature of the de Bruijn graph, the string abc is the one and only three-letter string that terminates at the node labelled abc so it follows that the part of the de Bruijn sequence of length four that ends with our traversing the arc *e* is *abcd*. Since this four-letter string occurs once and only once in the de Bruijn sequence, it follows that *e* cannot be traversed more than once during our walk. Therefore, for each de Bruijn sequence there is an Euler circuit whose label is the given de Bruijn sequence.

And the argument goes the other way: the label of an Euler circuit in our de Bruijn graph is a de Bruijn sequence: consider the binary sequence *S* (of length 16 if we work with Figure 6.3) that forms the label of the successive arcs in the circuit. Let *abcd* be a binary string of length 4 that crops up in *S*. Then the segment of the path labelled *abc* ends at the node with that same label and is followed by the arc *e* from the node *abc* labelled by *d*. Since this arc is only traversed once as we walk through the Euler circuit, it follows that the string *abcd* can occur no more than once in *S* because, each time we read the string *abcd* we are forced along the arc *e*. Therefore each four-letter binary string occurs no more than once in *S*. However, since the cyclic sequence *S* does contain 16 strings of length 4, it follows that each of the 16 strings of length 4 must occur exactly once in the cyclic list.

In short, each binary string of length 4 corresponds to an arc in the de Bruijn graph in such a way that a directed Euler circuit exactly matches the order in which the binary strings read in a de Bruijn sequence. We conclude that the label of an Euler circuit of a de Bruijn graph is a de Bruijn sequence and conversely any de Bruijn sequence arises as the label of an Euler circuit of the corresponding de Bruijn graph. We therefore have a complete theory that ties these two ideas together. Using the de Bruijn graph of Figure 6.4 you will be able to find de Bruijn sequences of length 32, a problem that would be very difficult without the de Bruijn graph to turn to.

Nets as machines

We may regard digraphs from a totally different perspective, that of an automaton. Mathematicians, computer scientists, and engineers all do this for their own purposes and some inkling as to why can be indicated here. One reason why mathematics is useful is that it leads to mechanical methods for solving problems. These methods were not necessarily performed by machines, but they could in principle be performed by inanimate devices. This allows us to solve problems, such as multiplying numbers together, without having to rediscover how to do it each time. When the mathematical recipe gets to the stage that it can be carried out in principle without thinking or indeed understanding what is going on, we say that the procedure has been reduced to an *algorithm*, a very ancient Arabic word that pervades modern scientific thinking.

As Mark Lawson explains in his book *Finite Automata*, it was no accident that as mathematicians were laying the foundations of the theory of algorithms, engineers were constructing real machines that implemented algorithms as programs. Algorithms and programs are just two sides of the same coin. However, some algorithms are much better than others. A good algorithm is fast and efficient. One approach to classifying the inherent simplicity of an algorithm is through *language theory*, and here the simplest algorithms are those that can be carried out on the simplest kind of machines, which are the finite automata.

The main ingredient of an *automaton* is a network in which the nodes are traditionally called *states*. Among the nodes there is an *initial state* and a number of *accepting* or *terminal* states. There may be more than one of these, and the initial state may also be an accepting state. I prefer the term accepting state to terminal state principally because these states are not final—in general an automaton can leave an accepting state under the action of further input letters. At any given moment an automaton A is in some state and may be acted on by an *input*, denoted by a letter of some set known as the *alphabet*, which has the effect of sending the automaton from one state to another. After a string of letters (a *word*) w acts acts on A, the automaton will either be in an accepting state or not, as the letters of w take the automaton through some succession of states. We say that a word w is *accepted* by the automaton, or is *recognized* by the automaton, if it leaves it in an accepting state. If not, w is rejected, and we say that the word is not part of the *language* recognized by the automaton.

If you are inclined to anthropomorphism, you can think of the states of A as *moods* with the accepting states representing the machine's good moods and the remaining states its bad moods. It wakes up in its initial state (which, like us, may be good or bad, depending on the particular machine's temperament) and the inputs to which it is subjected render it either in a good or bad mood. If it finishes in a good mood then it accepts the word, but if the word puts it in a bad mood, then it rejects it. The languages that we talk about in this context are not generally thought to be ordinary languages, although they are by no means excluded. Formal languages, taken in full generality, consist of arbitrary strings of symbols from some alphabet. Usually our automata are out to detect patterns, or the absence of them, within these strings, which are referred to as *words* despite not necessarily having a meaning in themselves.

For example, let us have a simple alphabet $A = \{a, b\}$. This will always suffice for our purposes, and indeed for most theoretical work; two letters are like two sexes, more than enough to create all the trouble you would ever want.

In the three automata of Figure 6.5, the initial state is labelled i and the accepting states are shaded. The arrows on the arcs indicate how a letter changes the automaton from one state to another.

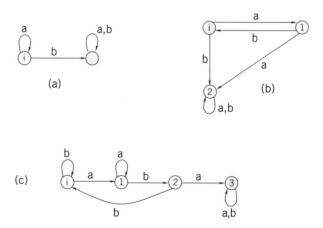

Figure 6.5 three automata

The automaton pictured in Figure 6.5(a) recognizes a word provided that it contains at least one instance of the letter b. A word consisting only of a's never takes the machine out of its initial state. Once the automaton sees a b it is happy, and it stays in its happy mood (the accepting state) no matter what it sees after that.

The machine of part (b) of the figure is not so easily pleased. This fellow will recognize a word only if it consists of a string on ab's, which includes the empty word (a string of zero ab's). (Quite generally, to say that an automaton accepts the empty word is tantamount to saying that its initial state is also an accepting state.) For example, the word *ababababab* will cause the machine in (b) to go from its initial state (which is its only accepting state) to state 1 and back again to i, four times. Since in this instance it finishes at the accepting state, it recognizes the word. However, as soon as it can tell it is not going to get a string of ab's, it moves to its sink state, 2, from which it will not budge. This will happen if you input a word beginning with a b, or if your word ever has two consecutive letters that are the same. Either of these events is enough to offend the machine as it will know that it is being offered a word that is not in its language, after which it totally loses interest. The word *ababa* would leave the automaton in state 1, which is still not an accepting state. State 1 is not a sink state, however—in this state the machine is still disposed to accepting the input string should it happen to continue in an acceptable way.

Before reading on, you might like to see if you can describe for yourself the language recognized by our third machine in (c). This automaton accepts the word *baababba*, but not *abba*. Closer examination will allow you to see that a word is in the language of this machine if it contains the *factor aba* and not otherwise. Indeed this is the smallest automaton that can be devised that accepts this particular language.

Having seen these few examples, readers may be inspired to experiment and design some automata of their own. Some examples to develop your skill would be automata that accepted the following languages: (1) words that contain *ba* as a factor; (2) words that contain both of the letters a and b at least once; (3) words that end in the letter a; (4) words of odd length; (5) words that begin and end in different letters.* You should bear in mind that your machine must accept the words described, *but no others*—your automaton must discriminate between acceptable words, and those that fail to qualify. After all, it is easy to make an automaton that accepts *all* words—just take the one state automaton where this initial state is also an accepting state with every letter taking that state to itself. We can input any word into this trivial machine and it will be accepted. The challenge is to create automata that accept desirable words, but are discerning enough to discard the rest.

You may like to dream up examples of your own, but you need to be wary—many languages that offer simple verbal descriptions are not recognizable by automata. For example, the language of all *palindromes* (words such as radar, minim, and redder that are themselves when spelt backwards) is not the language of any automaton. Any automaton that accepts every palindrome will be forced also to accept some words that are not palindromic.

It would seem then that automata are good at recognizing some patterns but not others. Automata also have some trouble counting, or to be more precise, there is always a limit to how many things any individual automaton can group together in pairs, and you will see why shortly. However, they can solve problems such as telling you whether a given number is a multiple of some particular integer. Indeed the machine of Figure 6.6 in effect does just that by recognizing multiples of 3.

In this case our alphabet consists of just the single letter, *a*. Clearly a string of *n* *a*'s, which we often write more conveniently as a^n, will leave this machine in its accepting state if, and only if, *n* is a multiple of 3. This automaton contrasts with previous examples in that in every case it needs to examine the input word in its entirety before it can decide whether or not that word is a member of its language.

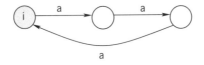

Figure 6.6 An automaton that counts in threes

However, any particular automaton cannot count past a certain number. (The automaton of Figure 6.6 will tell you if any given number is divisible by three, but it cannot tell you the outcome of the division as it completely loses track of how many times it has looped around the cycle of states.) In fact we can give an example of an unrecognizable language and demonstrate that this is the case. Interestingly, the argument makes use of the Pigeonhole Principle, introduced in Chapter 3.

Once again, let us revert to the standard two-letter alphabet, $\{a,b\}$ and let L be the language of all words of the form $a^n b^n$. That is to say, L consists of all the words $ab, aabb, aaabbb, \ldots$ that consist of a string of a's followed by an equal number of b's. Suppose that A were an automaton that recognized all the words of the above language L. This much is entirely possible, but we will show that A will be forced also to accept some words not of this type, so that the language of A is not L but rather it is some larger set of words.

For each particular number n, the word a^n takes A from its initial state to some state that we shall denote by s_n. Since $a^n b^n$ is accepted by A, the word b^n takes A from the state s_n to some accepting state, let us call it c_n.

Now since A has a limited number of states but there are infinitely many possibilities for n, it follows that there must be two different numbers, m and n say, such that the states s_m and s_n of A are the same, even though the numbers m and n are not. With this in mind, consider the word $a^m b^n$, which is *not* in L because $m \neq n$. This word is, however, accepted by A, as a^m takes A from the initial state i to $s_m = s_n$, and then b^n takes A from s_n to the same accepting state c_n, as before. As we explained above, it follows therefore that L is not the language of any automaton.

It is worth noting that the previous argument did use a version of the Pigeonhole Principle. With each number n, we associated a state of the automaton, which we called s_n. In this way we are assigning an infinite number of objects (all the counting numbers) to a finite collection of pigeonholes (the states of the automaton) and so at least one state has two different numbers, m and n, assigned to it; that is to say $s_m = s_n$. Although that is all we needed in our argument, we can of course say more—at least one state will have infinitely many different numbers assigned to it. This more general observation is the basis of the famous *Pumping Lemma* in automata theory, which says that once an automaton accepts a word of length at least as large as the number of states in the machine, then the automaton is forced to accept infinitely many words of a certain type, associated with cycles that arise when reading the input. This lemma is a tool by which it can be demonstrated that many languages are not recognized by automata, including the language of all palindromes and the language of all *squares*, which are words of the form $w^2 = ww$, where w is any string.

There are numerous applications of automata theory to theoretical computer science. Part of the reason it arises so often is that there are several equivalent ways of representing the class of recognizable languages and the different angles of approach are revealing. There are two other ways of introducing this class which, at first sight, bear no relationship with the machine viewpoint. The class coincides with the class of so-called *regular sets* (sometimes also known as the class of *rational languages*).

This collection is built up as follows from the letters of the underlying alphabet A, which are taken, by definition, to represent regular sets. If we have two regular sets, U and V, the set of words that results by pooling the two sets together is also regular (by definition, nothing to prove here). Mathematicians call this operation *union* and write $U \cup V$ for the set formed by taking the union. Similarly the set of words common to two regular sets is also regular—this is known as the *intersection* of the two sets and is denoted by $U \cap V$. Similarly we speak of the *product* of the sets U and V, denoted by UV, as the set that consists of any word from U followed (without a space) by any word from V, and this too is a regular set. In particular, the set V could be the same as the set U, in which case we write the product set as U^2, although this should in no way be confused with the familiar squaring operation that applies to numbers. Continuing this process we see that the sets $U^3, U^4, \ldots$ and so on of further products are also regular. Finally, the collection of words that result by taking the union of all the powers of a regular set, $U \cup U^2 \cup U^3 \cup \ldots$ is denoted by U^* and is also deemed, by definition, to be regular.

The class of all regular sets is said to be the collection of all 'languages', that is sets of words over the alphabet A, that can be constructed from A using these operations any number of times. For example, $A^*ab^2A^*$ represents the regular set that consists of all words containing the factor ab^2 somewhere in the string.

A famous theorem first proved by Stephen Kleene in 1956 then says that this collection is exactly the same as the set of languages that can be recognized by an automaton. A third characterization of the regular languages, that I will not explain here, is the class of languages that arise as inverse images of homomorphisms of the free semigroup over A.*

Suffice it to say that all three viewpoints have advantages. For example, look what happens if you consider any automaton and take its 'complement', in that we interchange the roles of all the accepting and non-accepting states. It is clear that a word will be accepted by this new machine if and only if it was *not* accepted by the original. In other words, this complementary automaton recognizes the complementary language of the original—the set of all words that were *not* accepted by the original machine. On the other hand, given automata that accept languages U and V respectively, it is not so obvious how to construct a machine that accepts the languages $U \cup V$, and another that accepts $U \cap V$. However, by Kleene's Theorem, it must be possible.

The algebraic approach to recognizable languages, briefly referred to above, is naturally symmetric and so lends itself to transparency when it comes to establishing results that are double-sided in nature. The machine approach however by contrast is extremely one-sided—an automaton has a unique out-edge from every state for each letter of the alphabet but there is no corresponding uniqueness as regards edges directed inwards. Despite the intuitive appeal of the machine idea, some symmetries of the theory may appear obscure if we insist on adhering to strict automata interpretation of all matters to do with recognizable languages. The algebraic approach makes it clear that given a regular language L, the language L^r of all reversed words is also regular. However, it is not immediately obvious how to build an automaton that accepts L^r given an automaton that accepts L, but again, by Kleene's Theorem, it must be possible.

The regular language approach is also a nice framework in which to show that certain operations on regular languages always yield more regular languages. One particular operation, known as the taking of *quotients*, is one of these. As you might imagine, the quotient of two languages is designed to be a kind of reverse to the product operation on two languages. Using the theory of quotients, we can show that there is a unique smallest automaton (that is to say, one with the fewest states) that recognizes a given language. This allows us a *syntactic* (that is to say, grammar-based) approach to building these minimal machines that are often used as pattern recognizers in real computers.

The 'mechanical' way of introducing regular languages may appear rather cumbersome. The definition of automaton has an arbitrary and non-symmetric feel to it. After all, why should the digraph always have just one initial state but any number of accepting states? Mathematicians would normally give that kind of clumsy definition short shrift. It turns out, however, that it makes no difference if we allow multiple initial states in that the language of such a machine is still regular and so can be recognized by an automaton with a unique initial state. (It does matter, however, if you insist on a unique terminal state—that would restrict the class of languages that were recognizable.)

Indeed, we can allow ourselves a lot of freedom in the definition of automata and the class of languages that are accepted does not alter at all. We can even allow the automata to be *non-deterministic*, meaning that when a letter acts on a state, the machine might move to a number of states or none at all. We can also allow for ε-actions where the machine may change states when you are not looking, under the action of the empty word, ε. It continues to make no difference in that the class of recognizable languages still coincides with the regular languages and goes no wider. There are great advantages in this. On the one hand, we can restrict ourselves to the standard definition of automaton without any loss of computing power. On the other, it is often easier for theoretical purposes to argue in terms of automata with more liberally defined features and so, if it suits us, we can indulge ourselves in that way.

The theory of regular and associated languages is a beautiful subject, replete with elegant constructions, and the starting point for it all is the study of certain kinds of digraphs. However, the most likely examples of automata that we are liable to meet in everyday life come in the form of vending machines or traffic light systems. These can be modelled as automata but with accessories as these machines also have outputs. A working vending machine, for instance, is always in one of a fixed number of states and responds to letters of an input alphabet in the form of customer selection options which stimulate it to move to a new state. In the process of transition there is also an output in the form of one or a string of available products. In a similar way, a traffic light system at a road intersection that responds to stimulation through road sensors is an automaton with outputs in the form of commands to traffic at the junction that it controls.

Automata with something to say

Computer science has an entire zoo of theoretical machines that can and cannot carry out certain tasks. The next step up from an ordinary automaton is a so-called *pushdown automaton* (PDA), which can be thought of as an automaton with a memory stack. Your access to the stack is limited, however, in that it is only possible to access the stack at the top. This is often likened to a pile of stacked dinner plates—to get to a plate in the middle you need to unstack the tower first.

This kind of machine does enjoy greater power of recognition than simple automata. For example, the languages of all palindromes, which is not regular, is recognized by a suitable PDA. The class of languages that PDAs accept is known as the *context-free languages*, a technical term derived from formal grammar theory which means, broadly speaking, that the meaning which can be derived from any particular formal grammar symbol is independent of the context in which the symbol may find itself. Although a wider class, the class of context-free languages lack some of the mathematical properties of regular languages, for example, the class is *not* closed under complementation. Moreover, some simply described languages, such as the language of all squares, are not context-free (nor regular). Those grammars not of the context-free type are, quite naturally, called *context-sensitive*. The heirarchy of grammars that correspond to various theoretical machine types is often known as the *Chomsky Heirarchy*, named after the famous mathematician, linguist, and political commentator Noam Chomsky.

The most general type of machine is named after the famous English mathematician and wartime codebreaker Alan Turing. A *Turing machine* consists of a tape, unbounded in both directions, and a programmable head that can move the tape in either direction, erasing cells and overwriting new symbols as it goes, depending on what it has just seen. Simple-minded as this sounds, *any* algorithm can be implemented on a Turing machine so this theoretical but simple construction potentially allows full exploration of what is and is not computationally possible.

In this book, however, we are concentrating on automata as they are the machines that come to us naturally as networks with nodes and edges decorated in one fashion or another. There are two standard types of automata with outputs known respectively as *Mealy* and *Moore machines*. In a Mealy machine the output is associated with the transition between states whereas in the Moore model the output is determined by the state itself. However, both models are equivalent in that any function that can be carried out by one of these automata types can also be performed by the other. In each case the machine can be fully described by a directed graph with appropriate labelling of states and arcs.

In Figure 6.7 we see two Mealy machines with very different purposes. The first example is reminiscent of the automaton in Figure 6.5(c), whose purpose was to detect the pattern *aba* in an input string. The Mealy machine is first and foremost an automaton that will pass through a series of states as dictated by the input string. However, as the machine executes a transition, it will print a binary symbol, either 0 or 1, in the manner indicated in the diagram. (Meaning that x/y indicates that y is the output accompanying the transition induced by input x.) For example, if the input were the string $w = bababab^2aba^2$ it would move through the series of

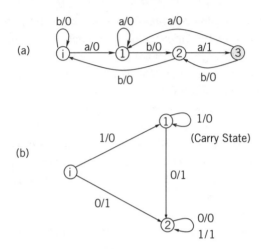

Figure 6.7 Mealy machines for two different purposes

states $i, i, 1, 2, 3, 2, 3, 2, i, 1, 2, 3, 1$. Since this is a Mealy machine, however, we are interested in its output, which in this case would be 000101000010.

What is this machine doing? Just regarding it as an automaton with accepting state 3, we observe that a word will be accepted by this underlying automaton exactly if it ends in the string *aba*. The Mealy machine always outputs a 0 except when it enters this accepting state when it registers a 1. What the Mealy machine is doing then is to *count* the number of times the factor *aba* occurs in the input string. The three instances of 1 in the output correspond to the three *aba* factors in the original input as you can now check (although the first two instances of *aba* in the input string overlap). Since an automaton with outputs can make a record of its transitions in this way, it can perform counting operations that are beyond the capabilities of a plain automaton.

The function of the second Mealy machine in Figure 6.7(b) is quite different for it adds 1 to a given number, when working in binary— we say that this machine *computes the successor function*. Since the operation of addition is carried out from right to left, we need to introduce the input string in reverse, reading it in from right to left. In the same way, if we write down the output string from left to right as it comes out of the machine, we need to reverse it to interpret the outcome as a binary number. Again an example is the best way to see what is going on.

In binary, the number 23 is 10111 (as this stands for $1 + 2 + 4 + 16$). Feeding this string into our machine (starting from the right) yields an output string 00011; the reverse of this string is 11000 = $8 + 16 = 24$, as we said it would be. The succession of states that the Mealy machine passes through as it reads its input are in this case $i, 1, 1, 1, 2, 2$.

If this looks a little perplexing, let us pause to see exactly what is happening. There are two things that can occur when we add 1 to a number written in binary. The simplest case is where the input number ends in 0, such as 1010. In this simple situation, all the machine needs to do is scrub out the final 0 and replace it by a 1, which is just what it will do: the machine passes from i to state 2 while doing this and then happily copies the rest of the string that it was given while remaining in state 2.

The more complicated alternative is when the final digit of the input string is a 1. In that case we need to change the final 1 to a 0 and 'carry' a 1 to the next column; the machine manages this by passing into the carry state, state 1. Here it will stay as long as necessary as it reads in the rest of the number: if the next digit is also a 1, that will have to be replaced by 0, and a 1 be carried to the next column. If that is what the machines sees, it stays in the carry state as it does this. It will remain there until it sees a 0, in which case it replaces that 0 with a 1 (because it still carrying a 1 from the previous step) and proceeds to state 2 from which point it merely needs to transcribe the rest of the string as it is given.

There is one slight blemish in the behaviour of this Mealy machine. By design, the length of the output string always equals that of the input. However, any number that is one less than a power of 2, such as 7 or 15, is represented in binary as a string of $1's$: $7 = 1 + 2 + 4$ is written 111 in base 2 and similarly 15 comes out as 1111. When we add 1 to such a number, its binary length increases by 1: $7 + 1 = 8$, which is written 1000 in binary, and similarly $16 = 15 + 1$ has the representation 10000. However, when these strings are fed into our Mealy machine the outcomes are 000 and

0000 respectively. In other words, the machine neglects putting the 1 on the front. However, provided that we interpret the output of a string of zeros as representing the binary number that begins with 1 followed by that zero string, then the machine is still telling us the right answer and the output always has an unambiguous meaning.

Automata with outputs are important in circuit design and can be generalized further to the class of so-called *rational transducers*. This leads into the fields of symbolic dynamics and coding with related applications that vary from the programming of computer compilers to designing compressed storage on compact discs.

Lattices

One of the more picturesque terms for a mathematical object is that of a lattice. Like family trees, lattices are networks directed from top to bottom but cycles in the underlying network are no longer forbidden. A *lattice* requires a little more structure than this, however. Each pair of nodes in a lattice must possess a *least upper bound* or *join* as well as a *greatest lower bound* or *meet*. If a and b are the two nodes in question we denote the join and meet of the pair by $a \vee b$ and $a \wedge b$ respectively. The expression 'least upper bound' is just about self-explanatory. We say that a node u is an *upper bound* of a node v if u lies above or is equal to v in the directed network. If this is the case we may indicate it by writing $v \leq u$ or equivalently, $u \geq v$. The least upper bound c of a and b is a node that is first and foremost an upper bound of both a and b. Moreover, c is the least of all these upper bounds: in other words if u is any node that lies above both a and b then we require that $c \leq u$ for c to qualify as the least upper bound of a and b. The dual notions of lower bound and greatest lower bound mirror that of upper bound and least upper bound. That is to say, we call u a lower bound of a if $u \leq a$ and we say c is the greatest lower bound of the pair a and b if c is a lower bound of them both and if u is any common lower bound of a and b then $u \leq c$. Since a and b enter symmetrically into the definition of meet and join it is clear that these operations are *commutative*, meaning that $a \wedge b = b \wedge a$ and $a \vee b = b \vee a$. (Note that if $a \leq b$ this causes no particular problem: in this case $a \wedge b = a$ and $a \vee b = b$.)

With these technicalities out of the way we can illustrate the concept of lattice through some natural examples. Lattices are by no means rare! A beautiful instance of a true lattice arises from the counting numbers $N = \{1, 2, 3, \ldots\}$. There is one node for each number and a lies below b if a is a factor of b, which we sometimes write as $a|b$. The greatest lower bound $a \wedge b$ of a and b is then the *highest common factor* (*hcf*) of the two numbers in question, while the least upper bound, $a \vee b$ is the *least common multiple* (*lcm*) of a and b. For example, if $a = 24$ and $b = 60$ then $a \wedge b = 12$ and $a \vee b = 120$.[1] The idea of meet and join of two nodes can be extended to three or more and indeed to any number of nodes. You might like to take up the challenge of finding the join of the set of the first ten counting numbers.*

And so you see that you studied lattices in school without ever knowing it. Many concepts of advanced mathematics have examples and motivation that lie at the most basic of levels and lattices are an instance of this. The lattice of divisors of the natural numbers has a common lower bound in the number 1 but no common upper bound. That is to say, there is no single number that is a multiple of every other number. Our lattice in this case is an infinite one, an infinite network if you like. To keep with the theme of finite networks and to give the opportunity of picturing a lattice of divisors, let us look at just a portion of this lattice, the sublattice of all divisors of the number 60. This lattice is pictured in Figure 6.8.

Although we have said what the nodes of this network are, we have yet to describe the arcs precisely. In this case it is best to talk of the edges being directed upwards so that if there is an arc from a to b then $a|b$, that is a is a factor of b. However, you will observe that $2|12$ also but there is no edge between these two numbers. This is because we only draw an edge from a up to b if there is no third number

---

[1] In case you have forgotten or were simply never taught, you find the hcf using Euclid's Algorithm, that is keep subtracting the smaller from the larger number until the two numbers in hand are the same. In this example you get, $60 - 24 = 36$, then $36 - 24 = 12$, and finally $24 - 12 = 12$, so that 12 is the hcf. The lcm is then $ab/\mathrm{hcf}(a,b)$, which in this case is $24 \times 60/12 = 2 \times 60 = 120$.

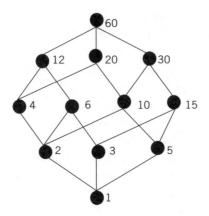

Figure 6.8 Divisor lattice of the number 60

c such that $a|c$ and $c|b$. It is still possible to tell from inspecting the lattice if one number divides another for the first is a factor of the second if and only if there is a path from the first up to the second. This works because the relationship 'is a factor of' is *transitive* meaning that if $a|b, b|c, c|d, \ldots$ and so on, then the first number in the list is a factor of the last. This behaviour is reminiscent of the ordinary $\leq$ relation on the integers. If transitivity could not be assumed we would need to indicate every relationship $a|b$ explicitly in the diagram otherwise we would lose information about which numbers are factors of which. Note that the layer immediately above 1 consists of the prime factors of the number at the top of the lattice.

The lattice of factors of a given number is a good example of a *partially ordered set*. There is an ordering on the set but some pairs of nodes are *incomparable* in that neither lies below the other. For example 10 and 15 are incomparable as neither divides the other. This contrasts with the usual linear ordering of the integers where given any two distinct integers, one always lies below the other. We do, however, require that a partial ordering have the property of transitivity, mentioned a moment ago, and also that it be *anti-symmetric*, meaning that if a and b are different nodes we cannot simultaneously have $a \leq b$ and $b \leq a$. This condition will ensure that we are never led around in a directed cycle in the network (even

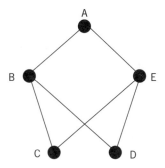

Figure 6.9 An ordered set that is not a lattice

though we may see cycles in the underlying undirected network, as we can in Figure 6.8).

In order to find the meets (hcf's) and joins (lcm's) of pairs of numbers you follow paths down and up respectively from the numbers in question and see where they first meet. For example, the lattice indicates that $12 \wedge 15 = 3$, while $4 \vee 15 = 60$.

The existence of meets and joins is by no means automatic in networks that represent orderings on sets. Figure 6.9 is a case in point. The partially ordered set defined by the network has no intrinsic significance. Rather it is introduced merely as an example of a (partially) ordered set that fails to qualify as a lattice.

Here we see an ordered set: $X \leq Y$ if there is a path from X up to Y in the diagram. For example, $C \leq A$ as we can pass from C up to A via E. The join of C and D does not exist! There are three upper bounds for the pair C and D. However, two of them, B and E, have equally good claim to being the join of C and D: they are both *minimal* upper bounds in that no other upper bound of the pair lies below them. However, neither can claim to be *the minimum* upper bound as B is not below E but neither is E below B. Similarly B and E have no greatest lower bound as both C and D are competing equally for the title. Worse still, C and D have no lower bound at all, so the pair C and D certainly lack a greatest lower bound. We see that the ordered set of Figure 6.9 is not a lattice. Although it is not quite obvious, it is simple to show that in any *finite* lattice any set of nodes has a greatest lower bound and a least upper bound. In particular this applies to the set of nodes of the entire lattice so that any finite lattice has a single absolutely least node (often denoted by 0) and a node that lies supreme above them all (denoted by 1).*

Lattices arise naturally right across crystallography and geometry but, more surprisingly perhaps, abstract algebra as well. Indeed lattices, with their two binary operations, are somewhat akin to ordinary algebra that is based on the familiar + and × operations of arithmetic. The prototype of a lattice in algebra is the lattice of all subsets of a set. The meet of two subsets is their intersection while their join corresponds to set union, the set that arises when we pool all elements of the two sets together. Often in modern algebra, a mathematician considers an object together with all of its subobjects of the same type or considers all the algebraic 'images' an object can have. This invariably leads to lattices of various kinds, often with very special properties, which in turn shed light on what goes on in the original algebraic setting.

Like other aspects of networks, lattices have emerged in recent years in their own right outside of mathematics. One area in which lattices have arisen is in the visual organization of complicated information. For example, the idea of *concept lattice* has become an important tool in many fields from linguistics to data mining, which is the extraction of deeper information from an enormous and seemingly unstructured mass of facts and measurements. Concept analysis is beyond the scope of this book but involves representing a concept both by its *intent* (properties of the concept) and its *extent* (the set of things that furnish instances of the concept). The lattice gives a picture of all the data simultaneously and so renders transparent relationships that may not be seen from a mere table of information. In particular (directed) paths in the lattice correspond to relationships between the concepts. The same information in tabular form would hide these structural features. The lattice structure can reveal attributes of the concepts involved. Moreover, new lattices can be formed, leading to interpretations of the subject matter that could not readily be arrived at by other routes.

7

Spanning Networks

We have talked about networks such as those of personal relationships and the World Wide Web that develop, if not of
their own accord, with a measure of autonomy, with the growth
largely determined by the will of individuals acting through their
nodes. However, a more concrete and traditional way that networks
arise is by overall design. Here the nodes represent places or people
that wish to form a network but the infrastructure of the edges
simply does not exist and has to be built. This can be a daunting
job.

What the engineers might have to hand is a paper network of possible connections between nodes. Each pencilled edge could be very
expensive to realize due to distance and other factors that separate
the nodes. All the same, it is easy enough to draw the possible edges
that might make up the network and to assign weights to each of
these edges proportional to the cost of making each edge a concrete
reality.

In the case of a road or communications network, it may not be
required or even desirable to put in place all the edges that might
be built. It may only be necessary to create enough edges so that
the network becomes one single component. The simplest way to
accomplish this is to find a *spanning tree* of the paper network. By
this we mean a tree containing all the nodes (and so just enough
edges so that the network consists of a single component). If we
had a network that was connected but was not a tree then it would
of necessity contain some cycles. An edge could be dropped from

each cycle and the network would still remain connected so that the simplest and cheapest network will be a spanning tree.

If all the edges were equally easy to build, so that all carried the same weight, so to speak, it would be very straightforward to design a suitable spanning tree: we would simply keep building edges until the network formed one connected block, mindful as we went never to add an edge that created a cycle, as that would be redundant. Although this would result in one of perhaps several possible solutions, every tree would be equally good as the number of edges of each spanning tree would necessarily be the same, that number being one less than the number of nodes in the network. Real-life networks are rarely so cooperative, however, as different edges have quite different associated costs.

We could just hope for the best and proceed in one of two simple-minded ways. We could build edges, one after another, to form a spanning tree by adjoining at each stage a new edge whose weight was as small as possible. This would see us successively building edges scattered all over the place. Eventually, however, given that the underlying paper network is connected, it would all come together and the outcome would be a spanning tree for our network. This seems a good common-sense approach but it is not clear that being greedy in this way—at each stage we always choose the cheapest option for the next step—will necessarily reward us with a spanning tree that is the least expensive. It is conceivable, for instance, that by choosing the second most expensive edge at some stage, we might avoid additional costs later on in the build.

A second approach, also quite naive, is motivated by the practical concern that it might be easiest to keep tacking on to what we have and build a larger and larger tree as we go. That is to say, as in the previous method, you begin with as cheap an edge as possible but at each stage you adjoin the least weighted edge that you can find to the tree you have already built. In other words you do not necessarily build the cheapest edge next, but rather the least costly edge that connects to the tree built so far. Once again, following this procedure, we shall eventually build a spanning tree for the network but it may look even more doubtful that it will be a tree of minimum possible weight.

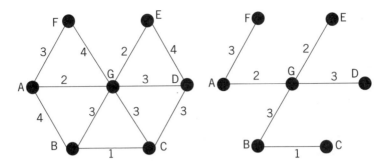

Figure 7.1 Network and minimal spanning tree

Fortunately, *both* these methods always work! The first is due to Kruskal while the second goes by the name of Prim's algorithm (sometimes equally attributed to Dijkstra but was first discovered by V. Jarnik in 1930). Each will provide us with a spanning tree of least weight.*

In the example of Figure 7.1 based on our Chinese Postman Problem of the previous chapter, we have a network and the given spanning tree of minimal weight can be found using either the approach of Kruskal or that of Prim. Taking on trust that these methods always give an optimal answer, we see that the weight of a minimal spanning tree in this example is 14 units.

Both these algorithms are known as *Greedy Algorithms* as at each step we maximize our gain at that step without regard as to how it might affect the overall result. When it works, greed is good, but in other similar problems greed can let you down, as we shall see later when we revisit travelling salesmen.

There happens to be a remarkable piece of mathematics due to Kirchhoff that allows us to calculate the precise number of spanning trees of a connected network. It is based on representing the network by an *incidence matrix*, which is a binary table that records which nodes are adjacent to which. The word *matrix* is a very fashionable one in modern management parlance and is often used there to indicate no more than a table of numbers. The word 'matrix' sounded snappy enough to become the title of the cult 1999 movie starring Keanu Reeves.

However, matrices come from mathematics and are of immense importance. Their power stems from the fact they can be *multiplied* in a way that is all their own. This leads to the enormous theory known as *linear algebra* that has grown over the last 150 years into one of the cornerstones of modern mathematics. It is a little fragment of matrix theory that offers a neat solution to this problem of counting spanning trees. Development of the theory would not be appropriate here as although it is not very deep, there are no shortcuts. However, for those who would like to dip into the topic, the Kirchhoff method as it applies to the problem of counting spanning trees is outlined in the final chapter.*

Sorting the traffic

In the previous chapter we considered the problem of making a two-way traffic system into a one-way set up and observed that this could only work if none of the streets of the network were 'bridges' in the network sense, that is to say edges that connected two otherwise disconnected blocks of the system. It was also stated that no other condition was required and, so as long as the network was free of bridges, a one-way system could be devised where it was possible to get from anywhere to anywhere else.

There are two approaches to solving this problem. The first is based on an alternative formulation of the condition that the network has no bridges, for this comes down to saying that every edge in the network is part of some circuit. Certainly a bridge cannot be part of a circuit for if it were, it would still be possible to get to the other side of the bridge by going the long way around the rest of the circuit. This is logically the same as saying that *if* an edge lies on some circuit, *then* the edge is not a bridge.[1]

And the converse is also true, for suppose that an edge e from a to b is *not* a bridge. That means that if we were to drop e from the

---

[1] Mathematicians are fond of this logical twist known as the *contrapositive*: P implies Q is logically the same as not Q implies not P. This takes a little thought—another example: *if we lose then we are not happy* is logically the same as *if we are happy then we didn't lose!*

network it would *not* split into distinct components, and so it must still be possible to drive from *a* to *b* somehow—this alternative trail from *a* to *b* together with the edge *e* will then constitute a circuit containing the given edge *e*. To reinforce this conclusion: an edge of a network lies on some circuit if and only if that edge is not a bridge. In consequence a network is free of bridges exactly when every edge is part of some circuit (the relevant circuit depends of course on the edge in question).

The circuit reformulation of the 'no bridges' condition allows us to explain how to build a one-way system in these circumstances. The method is very similar to the proof of the theorem on Euler circuits recorded in the final chapter. We begin with any circuit C_1 in the network, orienting each arc in the direction taken, we traverse the circuit until we return to the node *u* where the journey commenced. There may be additional streets between nodes in the circuit that are not part of C_1; put one-way directions on each of these in any way that seems convenient.

There may well remain edges and indeed nodes we have not visited but since the network is connected, in these circumstances there is an edge *e* from some node *v* that is *not* part of the original circuit C_1, to another node *w* that does lie in C_1. Find a circuit, C_2, that contains this edge *e*. Beginning at *v*, we put a one-way arrow from *v* to *w* and attempt to continue around the circuit C_2, leaving arrows in the direction we travel until we return to *v*. However, the circuit C_2 may meet up and share edges with the circuit C_1 that is already directed. When this happens, we do not argue with the orientations already provided, but respect the direction established earlier. In these circumstances we may have to take the long way around that part of C_1 to reach the end of the common segment of C_1 and C_2 before continuing our journey. The circuit C_2 therefore may not necessarily end up as a directed circuit but, using C_2 and C_1 together, it will be possible to drive from any point of C_2 or C_1 to any other point of C_2 or C_1, following the one-way system throughout. As before, after orienting C_2 in this manner there may be some additional streets in the system joining various nodes in C_1 or C_2—they can be oriented as we please, and may provide some useful shortcuts.

Continuing in this fashion, we eventually create a strongly connected digraph, that is to say a one-way traffic system that works. We can apply this method to the system of streets of our Chinese postman in Figure 7.1. The edge weights, which represent the lengths of the streets, are not relevant to the one-way system problem and so can be ignored. A nice circuit with which to begin is the 'ring road', C_1: $A \to B \to C \to D \to E \to G \to F \to A$ and we orient this cycle accordingly. Since this is an oriented Hamilton cycle, the network is already strongly connected and we can get away with orienting the remaining edges any way we choose. We can however continue to follow the recipe, noting that the edge GA leads into the circuit C_1. There is a circuit C_2 that contains this and all the remaining edges: C_2: $G \to A \to B \to G \to C \to D \to G$. Indeed C_2, consisting as it does of two separate cycles, is a true circuit and the given orientation is consistent with that of C_1 along the two arcs, AB and CD, that the two circuits share in common. All edges are now oriented and we have a strongly connected digraph, that is to say a one-way road system.

There is, however, a second way of approaching this problem that makes use of spanning trees and a version of Prim's algorithm. Since the network is connected, it is possible to find a spanning tree. We can do this by a *depth-first* search as follows.

Suppose that the network has n nodes in all. We choose any one of them, which we number 1, and then set out on a path through the network for as far as we can without repeating a node, numbering the nodes as we go, 1, 2, ..., k. Eventually we reach a dead end in that we can go no further without repeating a node. Whenever this happens we *backtrack* in the following way: we step back one edge along the path, and set out, if possible along a new path, numbering the nodes as we go $k + 1$, $k + 2$, ... until we are stuck, whereupon we repeat this backtracking procedure. Eventually this yields a spanning tree with all the nodes of the network carrying a number from 1 to n. The later a node is discovered by this search procedure, the higher will be its number.

Of course, sometimes after we have backtracked one step, no new path will be available from this node either and we have to backtrack again. Eventually, we may retreat all the way back to node number 1, from which point we will start off on a fresh path from 1, if some nodes remain unvisited.

This way of enumerating a spanning tree of a network is the key to the systematic search through a maze that we will meet again in Chapter 9. The pitfall to avoid is that of unknowingly slipping into a cycle. Given that the (connected) network is not a tree, then it will have cycles. An attempt to follow a path when constructing your spanning tree will then occasionally lead to a node that has been met before and we need to be aware of this. When this occurs, we should immediately backtrack and delete the edge we have just travelled upon from taking any further part in the construction of our tree as its inclusion would give a cycle.

If you have difficulty with picturing what is going on in a depth-first tree search, the best way to imagine the process is as a circumnavigation of an island. For example, if you leaf forward in the book to Figure 9.8(f) you will see a tree with labelled nodes (and edges). The purpose of this tree need not concern us at present but it serves as an example to search, beginning at the root node, that happens to be labelled 88, at the top of the picture. Imagine sailing around the boundary, setting off to the left as you sail down the page (so going to the right as we view the picture) and, clinging to the shore, listing each node that we meet until we complete our circumnavigation by returning to the root. We would be forced to backtrack a number of times and so some nodes would be repeatedly visited. Indeed the full list of nodes (together with repeats) that we meet in sailing around the tree is:

$$88, 37, 20, 37, 17, 37, 88, 51, 28, 51, 23, 12, 23, 11,$$
$$7, 11, 4, 11, 23, 51, 88$$

Deleting repeated nodes from the list, gives the following result for our depth-first search of the tree:

$$88, 37, 20, 17, 51, 28, 23, 12, 11, 7, 4.$$

Let us now see how to set up a one-way system using the tree we have built from our depth-first search. Our spanning tree having been found in this way, the rule we use for orienting the edges to yield a one-way system is very simple indeed. Take any edge e

that runs between the nodes i and j, say, with i the smaller of the two numbers. If e is part of our spanning tree, orient this arc $i \rightarrow j$; otherwise let the arrow point in the opposite direction. This orients the entire network in a manner that leaves it strongly connected.

We can apply this again to our example of Figure 7.1. Commencing at A, there is a single path that takes in every node in the network, which consists of the circuit C_1 above without the final edge that makes the path into a circuit. We number the nodes $A = 1$, $B = 2, \ldots, G = 6, F = 7$ as required to give our depth-first spanning tree. We orient all these edges consistent with this path; every other edge (i, j) is now directed from the higher to the lower of the two numbers i and j. This affords us a different solution to our search for a one-way system in that the arcs BG and DG of our first solution are reversed in our second.

The spanning-tree approach can be applied to any connected network in order to give it an orientation. If, however, the network has bridges then, inevitably, the resulting digraph will not be strongly connected. Because this algorithm can always be applied, it is by no means obvious that the spanning-tree method will always work and furnish a viable one-way system in any network that is free of bridges. To see why, and to see how the lack of bridges comes into it, you will need to read the details recorded in the final chapter.*

Recall that one definition of a tree is that of a connected *acyclic* network, which is to say one without cycles. The term *acyclic* can be applied to digraphs as well, meaning that the network has no *directed* cycles. Of course, any digraph whose underlying network is a tree will be acyclic but there will be other kinds as well: for example just take a triangle and orient two of the arcs away from their common node, orienting the final arc as you like it. This gives an example of a connected acyclic digraph whose underlying network is not a tree.

A simple characterization of acyclic digraphs is at hand all the same. Suppose that the digraph N has n nodes and suppose it were possible to name them $v_1, v_2, \ldots, v_n$ in such a way that arcs of the digraph pass only in the increasing direction of this order. That is to say that *if $v_i \rightarrow v_j$ were an arc of N, then $i < j$.* (We are not asking for the converse: just because $i < j$ we are not assuming there *is* an arc from i to j, just that there is *not* an arc from j to i.) It is then easy to see that N has to be acyclic, as the subscripts of the nodes we meet on any directed walk in N always increase and so we can never return whence we came, that is to say there can be no directed cycles in N.

Happily the reverse is also true: given that N is an acylic digraph, we can find an ordering of the nodes so that all arcs only lead upward in the order. The simple inductive proof is recorded in Chapter 10.*

Greedy salesmen

A little more can now be said of our travelling salesmen. Sometimes the salesman is replaced by a salesperson these days and, in the trade, the name of the problem is often shortened to the acronym TSP, an abbreviation that sidesteps any talk of gender specifics.[2] The sales representative has to start and end at home base having visited all of a designated set of cities. The challenge is to find the shortest route to go by. Traditionally the problem demands a cycle of the smallest possible total weight of the underlying network. If there is no Hamilton cycle, the problem of finding a Hamilton circuit of least weight arises.

The problem here then is to find a minimal spanning cycle as opposed to a minimal spanning tree. A greedy approach would be to follow the *Nearest Neighbour Algorithm*, where at each stage the salesman travels to the nearest available city not already visited and, when he has visited them all, goes home via the shortest possible route.

However, it is possible to find simple examples where the Nearest Neighbour Strategy is wasteful such as that of Figure 7.2.

Beginning at A, the Nearest Neighbour Strategy takes you through one of two equally good Hamilton cycles $ACBDA$, or $ACDBA$, each

---

[2] Somewhat ironically, the term *Traveling Salesman Problem* seems to have been coined in a 1955 paper by one of the most well known women mathematicians of the twentieth century, Julia Robinson, famous for her work on Hilbert's 10th Problem.

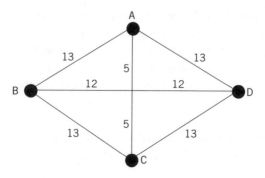

Figure 7.2 Near neighbour failure

of which has weight $10 + 13 + 24 + 13 = 60$ miles. However, $ABCDA$ is a better Hamilton cycle of weight $4 \times 13 = 52$ miles. Indeed, in this example, the short-sighted Nearest Neighbour Strategy fails in much the same fashion from whichever of the four towns the salesman uses as his base. What is more, this example could represent real distances on straight roads as the picture is based on four real and identical right-angled triangles of side lengths 5, 12, and 13, and the sum of the squares of the smaller of these three numbers equals the square of the largest, in accord with the age-old demands of Pythagoras on this matter, and so the diagram represents true separations.

In general, the TSP is an unsolved problem, as any problem leading in the direction of Hamilton cycles tends to be. Moreover, in many TSP networks, there is no Hamilton cycle and we have to make do with a Hamilton circuit. Since the TSP is a problem involving a finite search, it is strictly solvable as it is possible in principle to enumerate all the Hamilton circuits and find those of least weight. However, this is not in general a practical approach for real networks as the length of the calculation increases exponentially with the size of the network. A variety of usable algorithms are available and generally they aim to strike a balance between time of execution and optimality of solution. In general the architects of these methods seek to prove that their algorithm will always come very close to the best solution so that, in practice, their method will give you a near optimal solution in good time. They then measure the performance of their methods empirically against standard benchmarks (a network used by the French military is frequently cited as a standard test) and compare them with other commercially available software. The TSP differs from the Chinese Postman Problem of the previous chapter in that the postman is required to travel all the edges whereas the salesman only needs to visit the nodes and is not required to make use of every road available. In our solution of the CPP we implicitly assumed that we could always readily find the shortest path between any two points in a connected network. For small networks this can be easily done by inspection but what about the general problem? If we were reduced to checking every possible path between a given pair of nodes we would not really have solved the problem at all. The situation would be similar to that of the TSP which, although solvable in principle, was not solvable in practice, as a direct check of all possibilities leads to a prohibitive amount of work. However, this important subtext, the solution of the Shortest Path Problem, has been resolved.

Finding the quick route

A simple example suffices to show how this is done, even though, as in all simple examples, the answer may be got by inspection as there are few cases to consider. In fact it is instructive to stare at the picture (Figure 7.3) long enough to convince yourself you can see the answer before proceeding.

We shall show how the algorithm applies to find the shortest path from a to f in the network of Figure 7.3.

To be meaningful it would be best to consider the edge labels as *weights*, let us say costs as, in this instance, they could not represent direct physical distances because, for example, the net weight of two sides of the triangle cde is less than that of the third side (however, the numbers could represent lengths of winding roads). For that reason the interpretation of the solution would be the path of minimum cost, rather than least distance. The weights can, in general, take any positive value. For convenience we continue talking in terms of 'shortest paths'. The first to introduce a successful

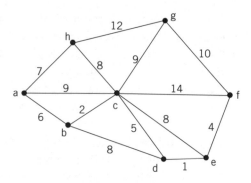

Figure 7.3 Finding the shortest path

algorithm for this problem was the Dutch mathematician Edsger Dijkstra (1930–2002).

It is a good algorithm from the computational point of view in that, if the network has n nodes, then the number of comparisons required is about n^2. One version of it runs as follows.

Keep in mind that our task is to find the length of a shortest path from a to f. The idea is to assign to each node a *temporary label* that represents an upper bound on the path length from a. At first, in order to err on the side of caution, all nodes are initially labelled ∞, which is a symbol taken to denote a quantity greater than any integer, apart from a itself that is labelled by 0. On each pass through the algorithm a temporary label, is replaced by a permanent one, which is the best we can do, in that it represents the length of the shortest path from a to that node. When you get to the stage where the node you are interested in reaching, f in our case, acquires its *permanent label*, we can stop. If we were to continue we would eventually find the shortest path length to all the nodes from our starting point a.

We shall write $w(e)$ for the weight of an edge e in the network. The rules of the algorithm are then as follows.

1. Set $v_1 = a$ and assign this node the *permanent label* 0. Assign every other node the temporary label ∞.
2. Repeat the following step as long as necessary:
    (a) Take the node v_i that has most recently acquired a permanent label, which is d, say. Look at each node v that is adjacent to v_i

but has not yet been permanently labelled and if $d + w(e) < t$, where t is the current label of v and e is the edge from v_i to v, change the temporary label to the (smaller) value $d + w(e)$.

(b) Having completed (a), take a node v that has the smallest temporary label among those still having temporary assignments, and make this temporary label permanent (if there are several of equal value, select whichever you prefer). Finally set $v = v_{i+1}$ and repeat step 2.

This process will eventually halt as each time we pass through the loop the number of permanent labels on the nodes increases by one. Importantly it does yield the length of the shortest path between the two vertices you declare an interest in. By following the procedure through on this little example, you will see how it works. In this case the shortest path is $a \rightarrow b \rightarrow c \rightarrow d \rightarrow e \rightarrow f$, which has total length $6 + 2 + 2 + 1 + 4 = 15$ units. In the course of working Dijkstra's algorithm you will find that c first has its temporary label changed to 9 before gaining the permanent label 8 on the next step, indicating that $a \rightarrow b \rightarrow c$ is the shortest path as far as c. Similarly the label sequence of g goes through the temporary 19 before acquiring the permanent 17, while f itself is temporarily labelled 22 before the optimal value of 15 is found.

As stated, the algorithm ends by providing us with the length of a shortest path, but not the path itself. It is a simple matter, however, to trace back and build the path: to keep track of the path we need only note whenever we assign a permanent label to a node, which node led to that label becoming permanent. This allows us to see, starting at f, where the shortest path has just come from, and in this way we follow the path right back to the beginning.

This algorithm is only one of a number. The Bellman–Ford algorithm also solves the same *single-source shortest path problem*. It has the advantage over the Dijkstra algorithm in that it can also deal with edges that have negative weightings. If the problem does not involve negative weights, however, the Dijkstra algorithm is the quicker of the two. The time complexity of the Bellman–Ford algorithm is proportional to the product ne of the number of nodes n and edges e in the network.

Another popular algorithm for this problem was developed by Floyd and Warshall. Although the time complexity of their algorithm is of order n^3, it nonetheless has several advantages. The amount of data stored at any stage in the process is quite modest and the algorithm simultaneously calculates the shortest distances between all pairs of nodes in the network.

Applications of algorithms like these are common in management science where large projects need to be organized efficiently and massive logistical mazes have to be negotiated in order to do this. In particular, the Shortest Path Algorithm can be applied in *Critical Path Analysis*. A typical example here involves a project that requires several stages. Some programs cannot begin before others have ended or, more generally, the cost and time taken by one part of the project depends on the overall stage of the project when the program is initiated. The problem is to finish the task in a way that minimizes time or cost. This leads to digraphs of the kind seen above and the required solution is represented by a *critical path*, which corresponds to a shortest path in the network from the initial to the terminal node.

Another name given to these problem types is PERT, which is an acronym standing for Project Evaluation and Review Techniques. PERT originated in the US Navy around 1956 while Critical Path Analysis was developed at about the same time by firms involved in commercial construction. Although there were some historical differences in the methods—PERT for instance incorporated a probabilistic element—both methods now come under the general heading of Project Scheduling Techniques and their applications are standard and widespread.

The P versus NP controversy

One of the greatest unsolved problems in all of mathematics is whether or not P = NP. Some say yes, some no, some say they have no idea, while still others suspect that the question can never be answered at all. This vexing question is all about exactly how difficult it is to solve a problem. Despite the fact that its precise formulation is a little technical, it can however be explained and, along the way, you will see that the TSP is not just a cute problem about saving time and money for travelling salesman but rather it represents one of the more fundamental difficulties of mathematics.

As we have mentioned more than once, an algorithm is a mechanical procedure that will solve a given type of problem. An algorithm may be complicated and proving that it works very difficult but, nonetheless, it might be quite easy to implement, at least with the aid of a computer. On the other hand some algorithms may be simple, even obvious, yet carrying them out can be prohibitively difficult.

The difficulty or costs associated with carrying out an algorithm are of two basic kinds. The first is *time* which is directly proportional to the number of basic computations that will need to be carried out in order to complete the calculation in question. The second might be called *space* by which we mean the amount of information that we need to store as the calculation proceeds. The P versus NP controversy centres around time, or more precisely the number of steps involved in an algorithm.

For example, suppose we are given *n* objects such as words or numbers and our task is to put them in order (alphabetical or numerical as the case may be). It is obvious that this *can* be done but how many steps will it take? Taking the example of number ranking, we ask: How many pairs of numbers might we end up having to compare before we are sure we have them all in ascending order?

In part this depends on what algorithm you use. One natural method is known as *bubble sort* where you make a number of successive passes through a list, each time swapping a consecutive pair of numbers if they are in the wrong order—in other words the bigger numbers rise like bubbles to the top. It turns out that the number of comparisons you may need to make is $\frac{1}{2}n(n-1)$ and we say that the *complexity* of the algorithm is of order n^2, denoted by $O(n^2)$, as, when the previous expression is multiplied out, n^2 is the highest power and, as *n* gets large, that is all that really matters, as it is the maximum power that dominates the long-term behaviour of this function.

However, there are better ways of sorting. The so-called *merge sort* algorithm uses a divide and conquer technique that is generally much quicker. The method involves repeated subdivision of the list into two, a process that can be represented as a binary tree followed by a merging of these lists which corresponds to a reflected image of the binary tree. The splitting process requires about $\log_2 n$ steps while the merging is a linear procedure in n so that merge sort has complexity $n \log_2 n$, which is known in the trade as *log linear*.[3] Since $\log n$ increases so much slower than n, this is a much smaller quantity than n^2. Indeed, since the log function increases extremely slowly, a log linear algorithm is considered to be about as good in practice as one of order n, a *linear* algorithm, which is very highly prized.

An example of a difficult-looking question that can be settled 'in linear time' is whether or not a network is a tree. Indeed a wider type of network that includes trees is the class of bipartite networks, which arise in the next chapter, and deciding whether or not a network is bipartite can be done in time of order n, where n is the number of edges. As mentioned above, the Shortest Path Algorithm is of order n^2 (in the nodes) while the standard method known as Gaussian elimination for solving a set of n linear equations in n unknowns, requires up to $\frac{1}{6}n^3$ basic arithmetic steps and so its complexity is of order n^3.

An algorithm that acts on a set of objects of size n is said to operate in *polynomial time* if its complexity is no more than n^k for some fixed power k, in which case we say the algorithm lies in the class P. As a rule of thumb, if an algorithm is in P it is regarded as being generally usable and otherwise not. This general position needs considerable amendment in practice. Some algorithms not in P are fine in that they will work quickly enough except for rare pathological cases while a polynomial time algorithm with a very high power of k might be useless. Indeed even a linear algorithm might be practically worthless if its true complexity in n was, for example, $10^{100}n$. It is all very well to say that the length of time taken to run the algorithm is

---

[3] The divide and conquer nature of merge sort is explained, for example, in K. H. Rosen's *Discrete Mathematics and its Applications*.

directly proportional to n but, even for small n, the number of steps would be billions of trillions!

There are some tasks for which it is known that a polynomial time algorithm is not possible. Examples include finding the best move for the board games chess and go. Algorithms that are not in P are sometimes loosely characterized as demanding *exponential time*, meaning that the number of steps required as a function of the n bits of input increases as least as fast as a^n for some constant a greater than 1. The point of this observation is that any such exponential function outstrips the growth of *any* polynomial function as n grows large and so exponential algorithms are, for sufficiently large data sets, not feasible to run, at least in the worst case. This apparent dichotomy between polynomial and exponential time is not quite the full picture for there is more to the story. There are processes of *intermediate growth*, functions that, in the long run, grow faster than every polynomial but slower than every exponential (for example, the function $2^{\sqrt{n}}$ is intermediate in nature), although you have to hunt hard to find algorithms leading to such growth. There are, however, some processes that are doubly exponential and there are monsters such as the reknowned *Ackermann's function* whose growth is so phenomenal that, although its values are 'computable', it transcends the ordinary laws of computability in a way that mere exponentials of exponentials never can.[4]

There are, however, a whole range of problems that do not look that hard but seem not to lie in the class P. The most simple to describe is the *subset sum problem*. We are given a finite list of integers and the task is to decide whether or not some subset of them sums to zero. (You can use any of the numbers you want once or not at all, but no repeats.) For example, suppose we are given the list

$$-131, -97, -90, -70, -35, -9, 1, 6, 7, 11, 18, 41, 50, 60, 78, 102$$

How can we decide this question? We want an algorithm, that is to say a list of instructions that will apply to *any* problem of this type, which will always settle the question, one way or the other.

---

[4] For those familiar with the technicalities, the Ackermann function is an example of a function that is recursive without being primitive recursive.

The obvious algorithm is to test every possible subset of numbers and see if each sums to zero or not. If we find one that does, then the answer is 'yes', if we exhaust all possiblities without ever finding a solution, the answer is 'no'.

This is fine in principle but unfortunately we are, in general, in for a full tree search, which is very bad news. You can see this by recasting what you are doing in terms of searching a binary tree as follows.

We can list all the members of any given finite set in some particular order: $a_1, a_2, \ldots, a_n$, say. A subset is then a list of some of these and any subset may be coded as a binary string of length n, that is to say a string in the two symbols 0 and 1. Given a subset, the first element of the string is 1 if a_1 lies in your collection and is 0 otherwise. We define the second, third, and so on symbols of the string in terms of the presence or absence of the corresponding a_i in your subset. For example, if $n = 4$ then the strings 0101, 1000, 1111 and 0000 would correspond to the respective subsets $\{a_2, a_4\}$, $\{a_1\}$, $\{a_1, a_2, a_3, a_4\}$, and the empty set respectively. (The empty set, often denoted by $\varnothing$ in books, is the set with no elements and is considered a subset of any set in order to keep the bookkeeping tidy.)

We can now see exactly how many subsets we have to check, for there is one subset for each binary string. The number of different binary strings of length n that can be built is $2 \times 2 \times 2 \times \ldots \times 2$, with n multiplications in all, as there are two choices (0 or 1) for each entry and each choice is made completely freely and is in no way dependent on any of the others. Hence there are 2^n subsets to check in our algorithm, which renders it *not* polynomial but rather an exponential process.

A *full complete binary tree* with n levels begins with a root and each node has two offspring down till we reach n levels below the root when we have a tree with 2^n leaves. (Our tribal allegiance tree of Figure 1.4 is an example where $n = 3$.) If we label the edges of the tree 0 or 1 according as the edge leans to the right or left then every path from the root to a leaf represents a unique binary string of length n, which, as we have just seen, can be interpreted as some subset of the original set of n objects. An example of such a tree appeared early in the first chapter. Figure 1.4 that we used to analyse our trio of tribesmen is a binary tree of height 3 (if we interpret T as 1 and L as 0) and each leaf represents one of the $2^3 = 8$ possibilities of tribal allegiance. In general, an algorithm that requires, at least in the worst case, a complete search of the leaves of a binary tree (or indeed an m-ary tree for any $m \geq 2$) is an algorithm that is not in the class P and so demands prohibitive computing time in large cases.

There is, however, one way in which the Subset Sum Problem is not bad at all. If some oracle were to give you a solution to the problem, we can readily check that it is telling the truth. For instance if I tell you a solution to our particular problem is the subset $\{-90, -70, -35, -9, 1, 6, 18, 41, 60, 78\}$ you can easily check, in polynomial time, that the negative and positive numbers in this subset both sum to the same number:

$$90 + 70 + 35 + 9 = 204 = 1 + 6 + 18 + 41 + 60 + 78$$

and therefore the entire set sums to zero. Therefore the answer to the Subset Sum Problem is, in this instance, 'yes'.

A decision problem such as the Subset Sum Problem that requires a yes or no answer and whose solutions can be *checked* in polynomial time is said to lie in the class NP. The abbreviation NP does not mean, as you might expect, Not Polynomial, but rather the more technical 'non-deterministic polynomial'. A problem is of class P if it can be solved by a so-called sequential *deterministic* machine while a problem is in NP if it can be solved using a *non-deterministic* sequential machine. All problems in P are automatically in NP. A big question in computer science, arguably the very biggest question, is whether or not the reverse is true. Are all problems in the class NP really in P as well? That is to say, is P = NP?

There is a one million dollar prize on offer waiting for anyone who can settle this vexing question one way or the other. Whether P (easy to do) is the same as NP (easy to check) is one of the seven problems that the Clay Institute of MIT list as worthy of this special *Millennium* status.[5]

---

[5] For rules and background, check the web page <http://www.claymath.org/millennium>.

The reason why this conundrum is particularly intriguing is that much which surrounds it is understood and has been examined extensively. Since the question is not settled, this means in particular that no one has ever actually proved that any standard NP problem such as our Subset Sum Problem absolutely does not succumb to a polynomial time algorithm. And you cannot assume these things are impossible just because many clever people have tried and failed. Only recently one particular problem that is very important in network security has been proved to be solvable in polynomial time, that of determining whether or not a number is prime. This was real news.[6] However, when it comes to the Subset Sum Problem, no one has found an approach that is a dramatic improvement on the naive tree search. That is, no polynomial time algorithm is known.

Furthermore, there are many important NP problems that have acquired special status. Both the Subset Sum and the Travelling Salesman Problems have been shown to be *NP-complete* and recently the same has been verified for the general problem of setting an $n \times n$ Sudoku puzzle.[7] This means that we know, for certain, that if we could solve the TSP in polynomial time then we could solve *every* NP problem in polynomial time and, in particular, the P = NP question would be settled once and for all in the affirmative.

The way this type of result is proved is diabolically cunning but is typical of the kind of thing mathematicians get up to all the time. The strategy in proving, for example, that the TSP is NP-complete is to show that, given *any* problem in NP, it is possible to reformulate it, in polynomial time, to a particular problem about the route of a travelling salesman. If we could settle the TSP in polynomial time, then the same would therefore apply to this arbitrary NP problem and so the TSP is NP-complete.

There is now a very extensive list of these NP-complete problems to choose from, including that of determining the presence of a

---

[6] Discovered in 2002, the AKS primality test has now been shown to have a very low power logarithm complexity: see <http://www.mathworld.wolfram.com/AKSPrimality Test.html>.

[7] This is not the same as the problem faced by the solver which is: given a partial sudoku grid that has a unique solution, find what it is. From what this author has read, the status of the complexity of this problem is less clear.

Hamiltonian cycle in a network. Indeed many on the ever-growing NP-complete list involve network isomorphism, network colouring, or the partition of numbers into sums. The hardest bit was getting going in the first place and this was achieved by Stephen Cook in 1971 when he proved that a certain problem of formal logic known as the Boolean Satisfiability Problem was NP-complete (although this terminology, now standard, was not used in the original paper).

Once we have one NP-complete problem the floodgates open, for we can show that another such as the TSP is likewise by transforming the known NP-complete problem into the new setting, thereby avoiding the necessity of showing that this can be done for *every* NP problem, which is what Cook, the originator of the theory, did have to do. Contrary to the expectations of the early days, networks are proving a natural source for some of the deepest, toughest, and most important problems in all of mathematics.

Going with the Flow

Network capacities and finding suitable boys

The two topics in the title of this section may sound a world apart but it transpires that the network problems involved are intimately connected. As is always the case, mathematics is indifferent to the applications we may have in mind and so serves to provide unexpected links between matters that would seem to be unrelated.

The first application is the very practical real-world problem of determining the maximum capacity of a given network. By contrast the second is that of finding out if it is possible or not to match up a group of girls with a given group of boys so that each of the girls is married off. The constraint that applies to this problem is that all marriages be by mutual consent, so a solution is not necessarily so easy to find.

To see an example of the first problem type we can return to the network of Figure 7.3 and interpret its meaning differently. Instead of distances, this time the weights on the edges represent capacities. Imagine the edges as pipelines of varying cross sections as given by the edge labels. Suppose then that our company wants to employ the network to pump oil from a to f and needs to know how much the system can take.

We can solve this problem by inspection, although even this relatively simple exercise requires enough thought to bring interesting features into play. There are three pipes leading from a of total capacity $7 + 9 + 6 = 22$, and so the capacity of the network cannot exceed this sum. However, can the network cope with this throughput or will bottlenecks arise?

The 7 units through pipe ah should not meet any obstruction as the pipes hg and gf have more than ample capacity to take this flow. Similarly the capacity 14 pipe from c to f should be able to take the 9 units arriving along ac with ease. However, let us see what will happen when we try a pump 6 units along ab. Upon reaching b we meet a node that can cope with a possible outflow of 10 units: up to 2 along bc and 8 along bd. Pumping 2 units through bc looks all right as cf has more than enough capacity to take these two units, even if it is already taking 9 from the pipe ac. However, at least 4 units will have to travel along bd but when that hits node d there is a problem. The total outward capacity at node d is $1 + 2 = 3$ and we have hit upon an unavoidable bottleneck. It seems then that we may only pump $2 + 3 = 5$ units through ab, which will then split into 2 units and 3 unit flows in completing the passage to f. The maximum capacity of this network is thus only $7 + 9 + (2 + 3) = 21$ units, and not the 22 we may have hoped for.

Having solved the problem, we can glean much by looking to the set of pipes that critically affected the overall result. The pipes that acted to limit the capacity of the network were ah, ac, bc, dc and de. This collection of pipes, taken all together, has one important property—if they were all removed from the network, it would fall apart, leaving no directed path of any kind from the source, a, to the sink, f. The name for such a collection of arcs whose deletion would cut the network's capacity to zero is a *cutset*.

Cutsets are a critical feature in the maximum flow problem as the total capacity of the network cannot possibly exceed the combined capacity of the arcs in any cutset, as every drop of oil has to pass through at least one of the pipes in the cutset. Therefore we already have an upper bound to the problem:

> *the maximum capacity of a network is no more than the minimum of the capacity of its cutsets.*

There are many other cutsets in this particular network: the three arcs from the source a is an example already mentioned of capacity

22 while the collection of arcs going into the sink f form another cutset, this time of capacity $10 + 14 + 4 = 28$. Neither of these, however, have the minimum measure of 21 that acts to limit the network's capacity to just this value.

A remarkable result, due to Ford and Fulkerson, is that this bound can always be attained. That is to say:

the maximum capacity of a directed network is equal to the minimum capacity of its cutsets.

We have already observed that the first quantity cannot exceed the second. The clever part of the proof is showing the reverse inequalilty allowing us to deduce the two numbers are always the same. It turns out that it takes only a few paragraphs of careful argument to establish this.* What is more, there is a workable algorithm, similar in feel to the Dijkstra algorithm for shortest paths but somewhat more complicated, that allows us to find the maximum capacity and actually set up an optimal flow. The idea is to begin with some feasible flow and augment it step-by-step. It can be proved that, by adhering to certain simple rules, a maximum flow can be attained.[1] It is not obvious that this would necessarily work as, if we go about it the wrong way, we can hit a dead end where we have a flow that is not optimal but which cannot be increased without first backtracking and reducing flows through some arcs. A workable algorithm that can operate in real time needs to avoid becoming a tree search of all possible flows in the network. It must navigate its way to an optimal result without excessive backtracking and searching through myriads of non-optimal cases that are of no interest, for only then can complex real-world problems be solved in practice.

As is often the case with networks, there are surprises in store. There are two reactions to a topic like the Maximum Flow Problem that are rather at odds with one another. On the one hand, some people approve of this kind of problem as they see it, quite correctly, as practical and important and so its solution represents a worthwhile mathematical direction justified on utilitarian grounds. On the other hand, others, especially those with no natural appetite

---

[1] See for example, *Introduction to Graph Theory*, p. 132, by Robin J. Wilson.

for engineering matters, are turned off by this setting and tend to lose interest, concluding that the subject has drifted into a direction that is all too mundane.

Happily, network theory manages to reconcile these two otherwise conflicting attitudes. The topic and the problems that it leads to such as job assignment problems are genuinely important and simply need to be solved. However, results such as the previous one are of real mathematical interest and when a development like that occurs it normally has consequences in unexpected directions that are revealed to the open-minded individuals who are prepared to follow the mathematical signposts they encounter.

The feature of real research that all mathematicians know but which is difficult to convey to the general public is that mathematics needs to be free to dip in and out of applications as the mood requires. A lot of good mathematics arises from practical problems. However, the mathematics that results often transcends the original problem and sheds light elsewhere, first in other parts of mathematics, and eventually in entirely different subjects. For example, problems in classical fluid dynamics took on a life of their own and the models involved then surfaced in economics. This is a natural development that ought not be hindered. We should not obstruct progress by insisting that researchers are driven only by narrowly defined goals announced in advance of the research program. Ploughing the same furrow, however important it seems, yields less and less as time goes by and if we ignore interesting diversions along the way on the grounds that they are not relevant to the matter at hand, it is a safe bet that we will be missing a trick. Real research is about the unknown and when an opportunity arises you need to drop everything and follow where it leads, despite what you may have written on the grant application.

At the level of teaching, it will always be the case however that people's appetite for studying a problem depends on the context in which it arises. Placing mathematics problems in enticing settings is a little art in itself. A question about footballers running around a pitch might be tackled better by boys than by girls, yet *the same problem*, when introduced in terms of people on a dance floor, encounters the opposite sex bias.

You can do things with the Max Flow Min Cut Theorem that have nothing to do with the capacity of grids and that you could hardly expect. One application is the resolution of the following problem that goes by the title of *Hall's Marriage Lemma*.

Marriage and other problems

Readers should not take offence at the original 1930s setting of this problem, as the distraction involved may lead to them missing the point entirely. Indeed I am going to dress the problem up with the opposite sex bias to that which it bore originally, as it seems more chivalrous.

Here is how it goes. We have a set of n girls that we would like to marry off and there is a set of m male suitors for our girls to choose from. You will be happy to hear that all marriages are to be by mutual consent. In other words, each suitor is only interested in some of the girls and not the others while, in just the same way, each girl is prepared to accept some of the suitors but not others, no matter what anyone says. The question is, given these constraints, can we marry every girl to a suitable boy?

The answer obviously is, 'it depends'. But what exactly does it depend on? That is the real question. Certainly it depends if nothing else on having enough boys in the first place—if there were more maidens than suitors, some of the girls would have to be disappointed. This simple observation brings home the fact that the problem is not a symmetric one. We have asked ourselves whether we can marry the girls off and are not worried about pleasing all the boys, something we should not lose sight of.

The basic observation of the previous paragraph can be taken one step further. Certainly there is no hope of solving the problem unless each girl has at least one suitor who she is prepared to wed. Taking this further, if we had two girls who only had eyes for the one man, then the problem is unsolvable as well. Between them there must be two suitors they are prepared to marry—if they each have one and only one sweetheart, it mustn't be the same fellow. And the same holds for any trio of girls: the collection of boys they are willing to take on must number at least three, as otherwise our matchmaking will end in frustration.

In general, then, if we take any set of k girls, the total set of boys they are (collectively) prepared to marry must number at least k if we are to see all our girls married. This is a necessary condition, known as *Hall's condition*, which must be satisfied if the Marriage Problem is to have any solution at all.

The pleasing aspect of the problem is that this is all we need—this necessary condition is also sufficient: as long as the set of suitable boys for each group of girls is not fewer in number than the girls, a solution can be found, a very happy outcome.

This pattern of solution we have seen on previous occasions when an obvious necessary condition has turned out to be enough to solve a problem. Just recently we saw that it is possible to have a particular flow through a network as long as the value of the flow does not exceed the capacity of any cutset. These similarities are not just coincidental but stem from the same root, for it is possible to view Hall's Lemma as a special case of the Max Flow Min Cut Theorem.

To see why, it is best to talk in terms of networks. The network of the Marriage Problem has two parts; we have one node for each girl and one for each boy and two nodes are joined if the pair represents a possible match. This kind of network is called *bipartite*. One particular bipartite network featured heavily in our discussion of planarity, that being the network $K_{3,3}$ first seen in Figure 3.5. This network consists of two sets of three nodes with each node in the first set adjacent to each node in the second set, and is called a *complete biparite network* for that reason. In general, a bipartite network consists of a pair of disjoint sets of vertices, G and B, and the edges of the network run exclusively between nodes of G and of B, with no edges between nodes in G, nor nodes in B. If every node in G is joined to every node in B, we have a complete bipartite network of which $K_{3,3}$ is a prime example (although there is no call for the sets G and B to have the same number of nodes in a complete bipartite network, as is the case in $K_{3,3}$). The Marriage Network is bipartite but would only be complete in the case where every girl and boy were happy to wed each other.

To say that a network is birpartite is the same as saying that it is 2-colourable, meaning that nodes can be coloured red and blue, say, in such a way that adjacent nodes have different colours. The blue and red nodes correspond exactly to the two 'parts', G (blue) and B (red), of the bipartite network. A more subtle characterization of bipartite networks, however, is as those in which all cycles have even length. This is quite simple to prove and leads to a method of deciding whether or not a given network is bipartite that is quick to implement.*

One consequence of this characterization that might strike you as surprising is that any tree is bipartite (as it has no cycles at all, and so satisfies the given condition without even trying). To 2-colour the nodes of your tree just start by colouring any endpoint red and then alternate the colours blue and red as you go. You can't go wrong!

The trick to reducing Hall's Lemma to a network flow problem is to take the bipartite network of the girls' and boys' marriage preferences and adjoin two new nodes, s and z, to act as source and sink respectively. We draw arcs from s to every node in G, the node set of the girls, and an arc from every boy in B to the sink z. All these new arcs are given a capacity of one unit.

Rather cunningly, we assign limitless capacity to each arc joining a girl to a suitable boy. Actually, as you will see, we do not need the capacity of their links to be infinite but rather some quantity q at least as large as n, the total number of girls.

For example, suppose that we have five girls and six boys with the marital preferences given by Figure 8.1. In this example, there are 6 boys for $n = 5$ girls but boy 6 is unwanted by all the girls (although, to be fair, he may be the fussy one) so any matching has to pair off five girls with five boys. Now b_3 is the only man for g_5, so there is no choice there. After that, b_5 is the only match left for g_2, and then fortunately the other three girls can be matched with the remaining boys to give our bridal pairs: $(g_1, b_1), (g_2, b_5), (g_3, b_2), (g_4, b_4), (g_5, b_3)$. Indeed you will be able to find one other solution as well.

The general connection has yet to be made between the bipartite network of the potential brides and grooms and the network flows in the augmented network. This is based on the observation that any

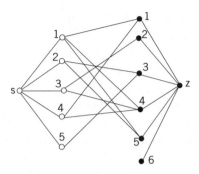

Figure 8.1 Flow network for a marriage problem

flow in the network corresponds to at least a partial matching of the couples for the following reason. In any flow from s to z, each arc from s to one of the female nodes carries either one unit or nothing. (Remember that the flow is 'quantized' and only comes in multiples of our basic unit.) Similarly an arc from one of the male nodes to z either carries one unit or nothing at all. Hence, despite the network having a very large outward capacity, flow of at most one unit exits from each female node, matching it with one and only one of the male nodes. In summary, in any flow through the network, each girl is matched by the flow with at most one (mutually) acceptable boy.

We are seeking to maximize the matchings of the girls which can be done by maximizing the flow. By the Max Flow Min Cut Theorem, we know the maximum value of the flow is the capacity of a minimum cut in the network. What remains to be demonstrated therefore is that if every set of k girls is joined collectively to at least k boys (for all $1 \leq k \leq n$), then any cut has capacity at least n, for that corresponds to a husband for each of our maidens.

To prove this, we assume to the contrary that despite Hall's condition being met, there is a set of arcs A that forms a cutset for the network and the capacity of A is less than n. We have deliberately constructed the network in order to ensure that no arc from G to B can be in A for the capacity of every one of these arcs is at least n, which exceeds the total capacity of A. Hence A consists of two sets of arcs, X and Y, with the arcs of X all emanating from the source, s, and leading into G, while those of Y all come from the boys' nodes B, and are directed to the sink, z.

Careful examination of this scenario leads to a contradiction, as you are about to read. The set X will have some number of nodes k, where k could be as low as 0 or as high as $n - 1$. The remaining set of girls number $n - k$ and, by Hall's condition that we are assuming is satisfied, they collectively have a set of at least $n - k$ boys to which they may be wed. For each of these boys, then, there is at least one girl, not in X, to whom he is linked. The arc from that boy to z must be in the cutset A for otherwise there would be a flow through the network via this girl and boy pair even after the arcs of A were dropped from the system. It follows that the size of Y is at least $n - k$ and so the total size of A, which is the sum of the sizes of X and Y, is at least $k + (n - k) = n$. This, however, contradicts our assumption that A had fewer than n arcs. All this proves that in the presence of Hall's condition, the size of a minimum cutset is at least n, and so the maximum flow is at least n. Clearly the flow can be no greater than n as there are only n arcs coming from the source, s, and each may carry no more than one unit. Therefore the maximum flow in this augmented Marriage Network is n which, as we explained above, corresponds to all of our n girls finding a husband.

This argument is a good example of an unexpected application of the topic of network flows. On the other hand you may still object that there should surely be some direct way of demonstrating the truth of the Hall Marriage Lemma without recourse to rather artificially embedding the set-up in a network flow. The answer is of course yes, and another short although subtle proof can be found in Chapter 10 that does not engineer the problem into a different setting.* At the same time we should also note additional advantages supplied by the flow model. Whether or not Hall's condition is satisfied, the flow model can be set up and will find an optimal solution, for a maximum flow corresponds to the matching of as many girls and boys as possible. In other words, the flow model will make the best of the situation at hand in all cases.

Applying one method to a variety of disparate problems is a good approach as not only does it show the generality of the method but it helps to bring out connections between mathematical topics that would not seem to be linked. It is often the case, and we saw it before with the Sylvester–Gallai Theorem of Chapter 4, that a problem is solved by recasting it in a new context. This happens quite often in mathematics where a problem in one realm is first solved in another. This can on some occasions rather annoy the experts from the first realm who are left feeling affronted. Thinking that they should have solved the problem themselves in the first place, they can then be galvanized into action and usually devise a proof that is more straightforward and natural, to them at least, with which they feel at home. Vanity, pride, and embarassment can be as strong as more noble motivations in discovering mathematical truth.

Harems, maximum flows, and other things

Short and succint as the statement of the Marriage Lemma is, the criterion is not an easy one to verify as it involves checking a condition for all subsets of the girls. In practice such assigment problems regularly fail to offer a complete solution and the flow model allows us to do the best we can, finding an optimal outcome through setting up a maximum flow. A situation that is met more often in practice than mass marriages is the assignment of jobs to workers. In this interpretation the set G represents jobs that need doing and the set B represents your set of potential workers. Each applicant is qualified for some jobs and not others and so we are left once again with the problem of trying to find a matching in a bipartite network, this time matching as many jobs to people as we can. The same network flow model that we used to prove the Marriage Lemma now serves as our model to find the best matching which will, once again, correspond to the maximum flow in the augmented network. In a way, this is not a different problem at all and the diagram of Figure 8.1 could well represent an example where we have five positions and six workers, although unfortunately applicant number 6 is not qualified for any of the jobs on offer and so is of no use to the employer.

The Max Flow Min Cut Theorem and the Marriage Lemma are both related to another problem type of real importance, that of finding disjoint paths between nodes in a network, an instance of which will surface when analysing *Instant Insanity* cubes in the next chapter. A good example to illustrate this problem type is the network of our Chinese postman, Figure 7.1. How many paths can you find in the picture that lead from A to D that have no edges in common? You should have little trouble finding a set of three: for example,

$$A \to G \to D, A \to F \to G \to E \to D, A \to B \to G \to C \to D.$$

At the same time this network has several cutsets of three edges, for example, removing the set of edges incident with A makes it impossible to travel from A to D. Do these two numbers, the minimum size of an edge cutset and the maximum size of a collection of edge disjoint paths between a pair of given nodes, always coincide?

The pattern we have seen with problems of this type persists in that it is clear that one of these numbers cannot exceed the other—any path between the two nodes of interest must include at least one edge from any given cutset and so it is impossible to have more edge disjoint paths than edges in any cutset. That the converse is true is not so obvious and was first brought to light by Karl Menger (1902–85) in 1927: the number of edge disjoint paths between two nodes in a connected network equals the mimimum size of an edge cutset for those nodes.

Like the Marriage Lemma, the difficult direction of Menger's Theorem may be proved directly or through use of a clever application of the Max Flow Min Cut Theorem. What is more, the Marriage Lemma can be inferred easily from Menger's Theorem, or at least from the nodal form of it, which we next explain.*

Instead of asking how many *edge disjoint* paths we can find, we may look to see how many *node disjoint* paths there are between two given nodes, meaning paths that have no nodes in common except for those at the beginning and end of the path. Any set of node disjoint paths will obviously have no edges in common either, but that is not necessarily true the other way round, for a pair of edge disjoint paths may well cross one another and so share a common node along the way. We see therefore that the maximum size of a set of node disjoint paths between two given nodes can be no more than the maximum size of a collection of paths that share no common edges.

For example, let us look again at Figure 7.1 and the nodes A and D. If we remove just the two nodes B and G (together with their incident edges) we see that A is now cut off from D. It follows that there can be no more than two paths in any set of node disjoint paths running between A and D and indeed it is easy to find a pair of paths that meet nowhere except at A and D.

Once again we have a pair of related numbers associated with the network: the size of a smallest *nodal cutset*, that is a set of nodes whose removal leaves our two given nodes disconnected, and the maximum possible size of a family of paths between the two nodes that share no common vertices. As with the edge form of Menger's Theorem, it is clear that the latter number is no more than the former, as any path between our two chosen nodes has to run through at least one member of any node cutset. The nodal form of Menger's Theorem (which is the version he originally proved) again says these numbers are one and the same every time: the maximum size of a set of node disjoint paths between a given pair of nodes in a network equals the minimum size of a nodal cutset between the pair. Good examples on which to try these ideas out are the network of the dodecahedron (Figure 3.8) and the Petersen graph of Figure 4.11: it is an instructive exercise to verify both the edge form and the nodal form of Menger's result for these networks for all possible choices of vertex pairs.

Contrary to the style of the presentation here, Hall's Marriage Problem has traditionally been introduced as a task of marrying off a set of boys with no necessity to marry all the girls. It is perhaps more natural to focus on the boys when considering the following generalization, known as the *Harem Problem*. In this variation we are looking to satisfy all the boys who each have expressed a wish for a harem of wives. That is to say, b_1 wants r_1 wives, b_2 wants r_2 wives, and so on where each of the r's is some non-negative integer. Indeed we may as well assume that all the r's are positive for if one of the boys doesn't want any wives then he is not part of the problem.

This is a generalization of the Marriage Problem as if we put all of the r's equal to 1, we return to the original setting where we know the answer. The reason for drawing your attention to this question is not because I have a particular interest in polygamy but because it is another example of use of the principle that was introduced earlier when we saw how we could extend the Euler circuit theorem for networks with no odd nodes, to the case where there are up to two odd nodes, by tweaking the solution of the simpler problem. Once again we can apply the principle by virtue of the following device.

We may reduce the Harem Problem to the Marriage Problem by imagining that boy b who required r wives is replaced by r copies of himself with each of the copies seeking just the one bride. Marrying this inflated set of boys off to the girls, one at a time, then corresponds to a solution of the Harem Problem and vice versa. Since we have now managed to work the problem into the original framework, we can state without further argument the conditions under which there is a solution: it will be possible for each of the boys to have the harem he desires if and only if it is the case that whenever we take any set of k of the boys, the collection of girls they are suited to marry is at least as great as the *sum* of all the numbers r for each boy in the set. For example, if $r_1 = 2$, $r_2 = 6$, and $r_3 = 1$ then the collection of potential wives for this particular set of three boys must number at least $2 + 6 + 1 = 9$ if we are to meet their requirements.

There is one particular set of circumstances that does automatically guarantee, however, that the condition of the Marriage Lemma is respected and moreover we can use it to settle a question that arose right back in Chapter 2 when we were talking about Sudoku grids. If there is some number r such that every one of the girls has exactly r boyfriends and every one of the boys has that same number r of girlfriends, then the conditions of the lemma are satisfied and the girls can all be married off.

In other words, we are seeking a matching in a regular bipartite network as every node, whether it lies in G or in B, has degree r. You might be wondering if in these circumstances the sizes of G and B must be the same. The answer is yes, and it is worth settling this before going any further. In any bipartite network, the sum of the degrees of G must equal the sum of the degrees of B, as all edges run between nodes of G and of B. In these circumstances this gives $rn = rm$, so that $n = m$ and there are equal numbers of girls and boys. Therefore, if we show that there is a matching in these circumstances, it must be what is naturally known as a *perfect matching* where each girl has a husband and each boy his bride.

It remains to check that Hall's Condition is met, so let us consider any set of k of the girls. These nodes are linked to let us say l nodes of B. The number of edges from this set of k nodes, that is to say the sum of their degrees, is rk. On the one hand, the number of edges coming from these l boys is rl, and on the other, this set of edges contains at least all the edges coming from the set of k girls, so we infer that $rl \geq rk$ and so $l \geq k$. That is to say, each set of k girls ($1 \leq k \leq n$) is suited to at least k boys, and so Hall's Criterion is met and a perfect matching can be effected.

Finally, as promised, we revisit and solve a problem we left unresolved in the second chapter. Recall that an $m \times n$ Latin rectangle is an array of m rows and n columns, with $m \leq n$, such that every row, which then has length n, contains each of the numbers $1, 2, \ldots, n$ (and so has no repeats) and no column has a repeated number either. The question was, is it true that a Latin rectangle can always be extended to make a Latin square where each number appears in each row and column?

The answer is yes, and the reassurance comes quickly through the Marriage Lemma. All that needs to be shown is that, if there are fewer rows than columns, we can always extend the Latin rectangle by one row. Carrying out this process $n - m$ times will then yield the full $n \times n$ Latin square that we seek.

We take for our set G the n columns of the rectangle and for our set B the n numbers 1 through to n. Each column is then linked with the set of $n - m$ numbers that do *not* appear in that column— our motivation for this is that we are trying to find a new number for each column in order to form the next row of the rectangle.

Let us take stock of our position: we have a bipartite network in which both parts, G and B, have n nodes and the degree of each node in G is $n - m$, which we shall call r. What is the degree of each node of B? Each of the numbers of B has appeared once in each row, and so has appeared in m rows and likewise in m of the columns as well, as no number appears twice in the one column. Therefore the degree of each node of B is also $r = n - m$, as that is the number of columns the number has *not* featured in to date. This means that we have precisely the situation just mentioned where we have a regular bipartite network in which every node has degree r. We now know that in these circumstances there is a perfect matching between the sets G and B, assigning the first column a new number, the second column a new number different from the first, and so on. Writing this down gives us the next row of our Latin rectangle, as every number is used and each column is extended by a new number.

As explained at the outset of the argument, we can now repeat this until we have filled out and created a full $n \times n$ Latin square. In conclusion, a Latin rectangle can always be padded out to form a full Latin quare.

9

Novel Applications of Nets

You have now seen enough to be shown some striking applications of network theory. The first is merely a puzzle, while the final one involves unravelling the mystery of life itself. As is often the case, the mathematics pays scant regard to our inital motivation and takes care of things in its own way.

Instant Insanity

This is a colourful and tactile Parker Brothers game that is based on a simple set of four plastic cubes each face carrying one the four colours, red, yellow, green, or blue, with each colour featuring at least once on every cube. The task is to form a tower from the four blocks so that each colour appears on each face of the four long sides of the tower.

Of course the problem is different for each different set of four cubes. With some colourings, the problem could have several solutions so that it may be relatively easy to stumble across one of them. On the other hand, with some colourings, there may be no way of doing it at all. The commercial toys are made to be pretty difficult and, for that reason, playing around with the cubes randomly can lead to real frustration, hence the name of this executive toy. There are a huge number of combinations you might try and it can be hard to find a systematic way of searching through them all to find the answer. In practice, coding up the problem in terms of related networks allows you to see your way through with relative ease. The nets organize the search for you in a manner that makes it possible to gain control over the problem and discover the solution with only a modest amount of trial and error.

It is easiest to show how it is all done through a good example. Rather than draw representations of three-dimensional objects on a flat sheet, which always involves some kind of distortion, we can resort to another device where we open up the geometrical object and fold it out flat. The flattened version is often known as a *net* of the original object, which is not to be confused with the way we use the words net and network throughout the rest of the book. The net of a cube is not unique in that a different picture can result if we separate along different edges. However, whatever version of the net we choose can be used to reassemble the cube from which it was first derived.

We open up each of the cubes to give us four cruciform nets with the sides labelled to identify the colourings involved (see Figure 9.1). We store all the information in the puzzle in the form of a network and then reinterpret the task in terms of features of that graph. In order to specify a network one needs to say what each node is and how each edge arises. In this case we draw one node for each colour and draw three edges for each cube, each labelled by the number of the cube. An edge is drawn between a pair of nodes representing two (not necessarily different) colours if these are the colours of a pair of *opposite* faces of the cube in question. For example, if we imagine reassembling cube 1 from its net, we see that the three opposite pairs of faces are (B, G), (B, Y), and (Y, Y) leading to edges labelled 1 running between the nodes B and G, B and Y, and a loop at the yellow node. We continue drawing the network in this way for each of the other cubes and the outcome is as seen in the figure. The network of the puzzle is not a simple network but will generally have multiple edges between pairs of nodes and some loops.

We have lost no information in passing from the collection of cubes to this single network, for once you know the colours of the three pairs of opposite faces of a cube, you can reconstruct the cube unambiguously. The question now is, what, in terms of the network, is the puzzle requiring from us?

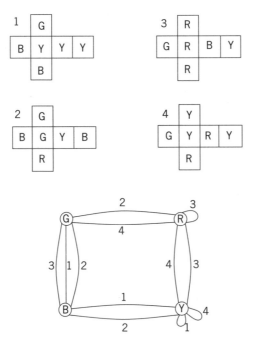

Figure 9.1 Network for Instant Insanity

If we can locate a Hamilton cycle in the network in which every type of numbered edge occurs once, we are halfway to solving the problem, for think what this represents. One cycle of this kind is to be seen in Figure 9.2(a).

Stack the cubes, one on top of each other, with cube 1 on the top, followed by cube 2, then 3, then 4. We can use the first cycle to arrange the tower so that the left- and right-hand sides have every colour. Start anywhere in the cycle, let us say at B, and trace a path around it, let's say clockwise. The first edge is labelled 1 so we may arrange cube 1 so that the left side is blue and the right is green. The next edge is labelled 4 so we adjust the bottom cube so that its left face is green and the right face is red; next we arrange cube 3 with red on the left-hand side and yellow on the right; finally we complete the circuit and in so doing place the second cube with yellow on the left and blue on the right.

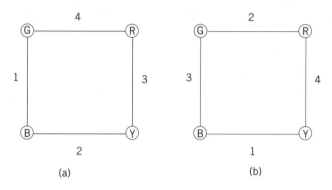

Figure 9.2 Hamilton cycles for Instant Insanity

By following this recipe we arrange it so that each colour in turn appears first on the left, then on the right of the tower of cubes. We want this to be part of our solution. That is to say we want to adjust the tower so that each colour also appears on the front and back faces while not upsetting the progress we have made thus far.

We can adjust the front and back of each cube if we wish by rotating the cube about the axis that runs through its left and right faces. In other words, we can swap the current front–back pair with the top–bottom pair for any cube while keeping the left and right faces the same. This measure of independence allows us to split the search for a solution into two similar parts for in this way we proceed by trying to repeat the process for the remaining front and back faces of the tower.

Having made this encouraging start, we look for another of these doubly Hamilton cycles—I say doubly Hamilton as not only is each colour represented exactly once through the nodes but we want every edge label to come up once as well. Since we wish to maintain the left and right faces as they are, we cannot use any of the edges of the cycle of Figure 9.2(a) again. The way to proceed, then, is to return to the original network of the problem, delete the edges of the cycle we have just used, and search for a fresh one with the same properties. In this example we are in luck and we see there is another and it is shown in Figure 9.2(b). We now use this to complete the puzzle.

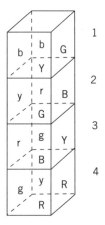

Figure 9.3 Solution to Insanity cubes

Once again, let us begin from the bottom left-hand node and read the instructions off clockwise. We rotate cube 3 as necessary so that its front face is blue and back is green, arrange cube 2 so that the front is green and the back red, then cube 4 with the red facing front and yellow at the back, and finally the top cube 1 with yellow facing forward and blue behind. This gives a solution to this particular Instant Insanity Puzzle as pictured in Figure 9.3. The lower case letters designate colourings of the left and back faces of the cube that would be hidden from the viewing angle in the picture.

Does the problem, then, always come down to finding a pair of these disjoint doubly Hamilton cycles in the underlying network? Well, no: there is more to it as the solutions are not always in that form.

For example, suppose that for a given set, one cube had a pair of opposite faces coloured blue, another a red pair, a third a yellow pair, and the final cube had a green pair of opposing faces. Lining these pairs up in our tower would immediately solve half the problem for us, yet the corresponding edges do not form a cycle in the network. Indeed these colourings manifest themselves in the network as four loops, one at each colour node, labelled by the four different edge numbers 1 through to 4. However, this will work just as well as the first ingredient to our solution.

What is required in order to find a suitable position of a pair of parallel sides of the tower (left–right or front–back) is a set of four edges of the network, carrying each of the four labels 1 to 4, that have the property that every node (i.e. colour) occurs once and every node has degree 2. A regular network of degree 2 such as this is not necessarily a four-cycle but could consist of a three-cycle and a loop at the remaining node, or a pair of two-cycles (multiple edges), or one two-cycle and a pair of loops, or finally, as we have already mentioned, four loops, one at each node. Any subnetwork of this type is known as a *factor* of the original network and to complete the problem we need not just one such factor, but two. Moreover, these factors must be *edge-disjoint*, that is to say share no common edge.

Any such pair of factors leads to a solution of the Insanity cube problem and conversely, if you find a solution then the colourings of the left–right faces and the front–back faces give you an edge-disjoint pair of factors of the full network of the problem with all the stated properties. It follows that all solutions can be found through finding all such factor pairs. The method then is to find one factor, and then look for another, edge-disjoint to your first choice. If there are none to be found, try beginning with another factor and repeat the process. In this way, all solutions are found quite readily and, if the problem is unsolvable, that will be revealed too

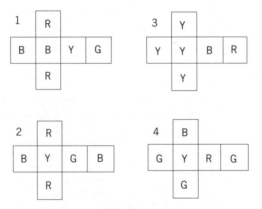

Figure 9.4 More Insanity

as you conduct a full search for the edge-disjoint factor pairs of the network. As another example, in which the factors are not just cycles, you may care to solve the Insanity problem for the previous set of four cubes in Figure 9.4, given by their nets. The solution is in the final chapter.*

Sharing the wine

Simeon Poisson was one of the greatest mathematicians of the nineteenth century. However, the young Frenchman took some time to find his calling, having spent his early years failing at one profession then another. The clarion call to mathematics apparently came about in an innocent way when he discovered that he had little trouble solving a puzzle of the following kind while others around him got in a muddle. It is a juggling and pouring problem.

Two friends wish to share equally eight litres of wine that fills a big pitcher. They have at hand two empty vessels of capacities five and three litres respectively. Can they manage the task of creating two four-litre portions?

It is implicit in the problem that all the friends are allowed to do is to pour wine from one jug to another until the receiving jug is full or the pouring jug is empty. Although it takes seven steps, the task can be managed and you are welcome to try to see your way through to the answer. There is a general method however that allows you to find all possibilities by means of a tree search organized on an ordinary xy-grid. At any one time the wine is divided into three portions with values x, y, and z say, where x is the amount in the five-litre jug, y the amount in the three, and z is the volume of wine left in the original large container. Since these three numbers always sum to 8, we only need keep track of the first two co-ordinates, x and y, to know the exact whereabouts of all the wine. We can therefore systematically plot all possibilities by starting at the initial position of the puzzle $(0, 0)$ and drawing arrows from each point arrived at to each possible new position.

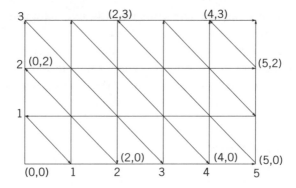

Figure 9.5 Sharing the wine

The plot of all possibilities is quite constrained; we know that x and y are always integral and never less than zero, and indeed the following hold:

$$0 \leq x \leq 5,\ 0 \leq y \leq 3 \text{ and } 0 \leq x + y \leq 8.$$

From any point, the puzzle may head off in at most three directions: if we pour from the small container (y) to the large one or the reverse, the arrow moves parallel to the y-axis, similarly we move parallel to the x-axis should we pour between the medium container and the biggest jug, while if we pour between the two smaller containers we move at an angle of $45°$ to the axes parallel to the line $x + y = 8$. The complete description of all outcomes can now be drawn and is to be seen in Figure 9.5.

In order to ensure we do not miss any possibilities, we begin from the root of the tree (0, 0) and draw arcs to each feasible outcome of the first step; from the origin we can get to the points (5, 0), (0, 3) only, giving us all paths of length 1. We then mark the inspected node (0, 0) with a star, as we have located all of its offspring. We continue in this way searching through the tree for any *new* possibilities and follow down each branch until no new offspring occur. We must, however, avoid cycles: if two paths meet up to form a cycle, it is not necessary to include the final directed edge of the longer path as that node is reachable using a shorter path not involving that arc. If two paths of equal length from the root meet, we can safely ignore the final arc of one of them—that arc might provide us with an alternative equally good solution but not a better one.

In this problem we are actually on the lookout for the node (4, 0) as then we have divided the wine into two four-litre portions in the two larger pitchers. Indeed we discover that this is feasible, and there are two routes to the solution:

$$(0, 0) \to (5, 0) \to (2, 3) \to (2, 0) \to (0, 2) \to (5, 2) \to (4, 3) \to (4, 0)$$

or

$$(0, 0) \to (0, 3) \to (3, 0) \to (3, 3) \to (5, 1) \to (0, 1) \to$$
$$(1, 0) \to (1, 3) \to (4, 0)$$

but the latter takes one more step.

If you care to try your hand at a similar problem, suppose we have three jugs of capacities 10, 7, and 4 litres, with the largest one full. Your job this time is to find a way of getting exactly two litres of wine in one of the pitchers.*

In this example we have used what is known as a *breadth-first search* of the tree where we begin at the root and determine all arcs leaving each node and so find the offspring of each node. This contrasts with the *depth-first search* that we have used on a number of previous problems such as our spanning tree for generating a one-way traffic system and indeed the original Knaves and Knights questions of the first chapter.

In a depth-first approach we search for a solution by beginning at the root but build a path as far as possible down the tree until a leaf is reached. If none is found, we backtrack one arc and proceed down in another direction from the last fork. If no solutions are forthcoming we may be forced to backtrack more than one step and in this way are led through a systematic search of the entire tree, looking for solutions.

The best search method to use in part depends on the requirements of the problem. If the tree of possibilities is large, the breadth-first method can become very unwieldy and the backtrack method, which traces only one path at a time, is the best way to go. If, however, we are keen on solutions with short paths, the breadth-first approach, which moves from shorter paths to longer ones, is the more likely to find shorter solutions quicker.

Jealousy problems

This represents a range of problems that arise when a task needs to be carried out when certain people, animals, or objects involved cannot be trusted alone together. Often framed in the form of jealous wives and husbands, the original eighth-century version is due to Bishop Alcuin of York and so arguably represents the oldest of network problems. We are invited to imagine a small boat owned by a boy who is left with the responsibility of transporting a fox, a goose, and a bag of corn from one river bank to the other. The trouble is that his boat is only big enough for himself and one of his three possessions and to make matters worse, the fox will attack the goose and the goose will eat the corn if left unsupervised. What is he to do?

He has to be patient—it will take him seven river crossings to do the job safely although there are two equally good ways of going about it. We can analyse all possibilities by a digraph (Figure 9.6) where each node describes the current state the job is in. For example our root is labelled $FGB^*/$, indicating that the Fox, Goose, and Bag of corn are together on the near shore with the * indicating the position of the boat (and hence our little boy). From each node we direct an arc representing a course of action, labelled by the object transported, or blank if the boy is crossing the river alone. Any arc that leads to the pairs FG or GB together without the protection of their guiding star is forbidden and so not drawn.

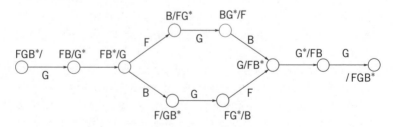

Figure 9.6 Fox, Goose, and Bag of corn problem

The digraph of all possibilities is shown in Figure 9.6, showing that there are just two equally good ways he can go about it. Indeed each solution may be obtained from the other through interchanging the symbols B and F throughout, as they enter the problem symmetrically (although not from the viewpoint of the goose!). The goose, the middle object in the pecking order, is the most liable to eat or be eaten and she has to be guarded carefully throughout.

There are other similar and more complicated problems that have been spawned by this, the classic example. Sometimes three married couples have to get themselves across the river without the risk of infidelity, while on other occasions there is a mixed group of missionaries and cannibals where, as you can imagine, it is the missionaries who have to keep their wits about them if everyone is to reach the other side in one piece.

Mazes and labyrinths

The Cretan maze is perhaps the oldest network in the world. Its design was found on a clay tablet in the ruins of the palace of King Nestor in Pylos, in western Greece, and dates to around 1200 BC. For millennia it seems to have functioned as the universal design of a basic maze for depictions of it, with remarkably little variation, are to be found not only in ancient Greece, in the ruins of the Italian city of Pompeii, and the floor of Chartres Cathedral in France, but outside Europe and the Middle East in carvings in Peru and aboriginal art in Australia.

The design itself is not so simple, yet it can be built up using a simple rule which perhaps accounts for its continual discovery and rediscovery across the ages. Figure 9.7 gives a picture of the maze itself. We begin with a cross, four L-shapes, and four dots as shown. The procedure is to join up the ends of the figure symmetrically, beginning as shown with a central pair, whether they be ends of lines or dots, always joining the next two ends, one from either side. Since there are sixteen endpoints at the beginning of the process, the maze is drawn in eight steps. In (a) we see the intial configuration and then the process is shown in (b) after two steps, in (c) after five,

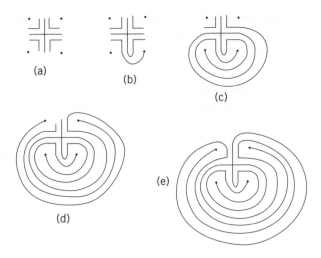

Figure 9.7 The maze of the minotaur

in (d) after seven, and the final picture is the complete maze after all eight arcs have been drawn.

Legend has it (there are many variations), that the Cretan maze was the lair of the minotaur and Dionysius thwarted the beast and solved the riddle of the labyrinth by use of Ariadne's thread. His lover Ariadne gifted him a sword and a golden thread to track his way through the maze and make good his escape after killing the minotaur that had captured an Athenian youth.

Confusing as it may appear with its layered paths, this labyrinth offers only one route from the outside to the centre so there is no opportunity to become truly lost. This differs from some other real mazes such as the famous example in the grounds of Hampton Court in Richmond, England where thousand of visitors each year are happy to get themselves lost and disoriented as they try to find their way back out. Although easy to negotiate once the plan is revealed, it is less easy to get the overall picture when you are wandering between the high hedges of this beautiful trap.

One rule of thumb for a systematic search of a maze is to follow the left (or right) wall throughout. This will allow you to create a map of the maze and so reveal all its secrets, including that of escape.

As with any search of this type, the key is to avoid an infinite loop. You may need help from Ariadne in one guise or another in order to recognize when you have completed some kind of circuit within the labyrinth. By following this strategy you are in reality revealing the underlying network of the maze and finding a spanning tree of that network, N. To construct N, we draw one node at each junction in the maze where a choice of path is offered and join one node to another if there is a direct path from the first node to the second in which we do not meet any other nodes along the way. However long this path may be as it winds around the maze, it is still represented by a single edge.

To conquer the maze, you do not necessarily need to find all of its edges, for a spanning tree will do the job by providing you with a unique path between any pair of nodes in the maze. The spanning tree can be systematically constructed using just the same enumerative approach that we introduced in a preceding chapter when using a spanning tree to solve the one-way traffic problem. As always, and this is the real trap when dealing with mazes, the pitfall to avoid is that of being led around in circles. This can be achieved by backtracking as soon as a path leads to a node already visited, so you need to be able to recognize when you have seen a node before, which can be tricky in real mazes where the designers deliberately make many of the junctions appear identical just to confuse those who venture in.

The designers also purposely make the exit node hard to find. Often a maze has a *centre*, which is a junction with many paths leading in and out and the one and only path leading to the exit goes via the centre. A visitor to the maze will soon find him- or herself in the centre from which they will find it difficult to escape, for if they wander about in a random way they will meet many paths that will lead them back whence they came, even though the path may look tempting by leading off in a new direction, only to wind back on itself and return then to the middle of the maze. Maze designers sometimes speak of the 'valving' being against you when most paths from a certain spot flatter to deceive and are there in order to lead you everywhere within the maze except the one exit you are striving to find. Of course by making the maze complicated enough it would be possible to make it near escape-proof. The art in real maze design is to create a maze that, although based on a network of modest size, still manages gently to confuse those who drift around it by tempting them along paths leading to nowhere in particular.

Constructing a spanning tree, either mentally (which can be hard) or on paper, will help, but return trips to the centre may still be unavoidable. However, patient building of the tree will at least ensure that even if you do revisit certain nodes, you will not reuse edges, so eventually the way out will be found. If you are lucky, you may escape without having to map out an entire spanning tree but a cunning design will ensure that even the most careful of visitors will most likely have to do a lot of walking before she finds her way out of the labyrinth.

Trees and codes

The topic of codes and ciphers is one of the hottest in modern applied mathematics and seems likely to remain so for some time to come. The most used piece of software in the world is the so-called RSA program that safely encrypts personal and other data over the internet. The development of these so-called public key cryptosystems, which had seemed impossible to create only thirty years ago, has made viable the commercial use of the internet and its World Wide Web.

However, coding is not always a matter of secrecy and often the purpose of a code is simply to store information in a succinct and usable way. One of the most basic of ideas is that of a *prefix code* where objects are encoded as strings of symbols in such a way that no string forms the prefix of any other. One standard example of this is a telephone directory in which a telephone outlet is identified as a number. It is vital that no number a is simply the beginning of some longer number b as, if that were to happen, it would never be possible to phone b at all as anyone dialling the number would be put through to a first!

There are many ways of forming prefix codes, nonetheless. Perhaps the simplest is to make sure that all the code words have the same length so that none can form an initial segment of any other. Consider the following common problem. Suppose that we wish to develop a way to represent the letters of the alphabet using binary strings of 0's and 1's. Since there are 26 letters and there are 32 possible binary strings of length 5, we can encode the letters of the alphabet using strings of that length. However, in English, or indeed any language, not all letters occur with the same frequency—far from it. Consequently, it would be more efficient to use binary sequences of different lengths, with the more frequently occurring letters (such as e, i, and t) represented by short strings.

Practical considerations of this nature undoubtedly went into the design of the Morse code in the nineteenth century, which was the basis of the first instantaneous electronic transfer of information across the world. However, if there are many symbols, such as 26 alphabet letters, a trial-and-error method for constructing the tree of possibilities is not efficient. Indeed, if we decide to include more grammatical symbols such as punctuation, spacing, upper and lower case, and so on, the number of symbols we wish to represent increases considerably. However, an elegant method due to David Huffman provides a technique for dealing with this question through the construction of a certain tree.

We give here only a simple example but it is enough to show how to go about it and why it works so well. Our task is to construct a prefix code for the six letters a, o, q, u, y, z that occur in a sample with respective relative frequencies 20, 28, 4, 17, 12, 7. The following natural construction builds a tree from which suitable code words for each letter can be read. The beauty of the resulting prefix code is that it is optimal, meaning that it will allow the translation of the passage as the shortest possible coded binary string. Any other coding of these six letters as a prefix code would yield a coded passage that was as long or probably longer than the one provided through the Huffman code. The procedure is carried out in full in Figure 9.8.

We begin by forming a list of nodes, six in this case, one for each symbol to be coded up, and label each node by its relative frequency

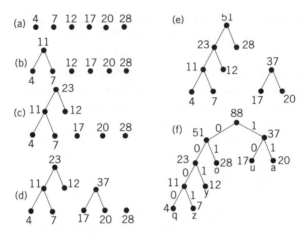

Figure 9.8 Building a Huffman tree code

in the passage and list them in order of increasing frequency. These nodes are to be viewed as six single-node trees, that node being the tree's root. At each stage we combine two of the remaining trees into a single one by taking the pair whose roots are labelled by the lowest numbers. The two trees are combined under the umbrella of a new common root that is labelled by the sum of the labels of the roots of the two trees in question. This is continued until we have a single-rooted tree. The number of steps required for this formation is one fewer than the number of symbols to be coded, so there are six stages in all from start to finish in this little example. (See Figure 9.8.)

Once the tree has been built, the required code can be read off as follows. Label the edges of the tree either 0 or 1 according as the edge goes to the left or the right in the tree. Label the leaf nodes with the letters that correspond to the frequencies indicated by each node. The code string for each letter is then read as the sequence of edge labels from the root of the Huffman tree to the corresponding leaf. In this illustrative example we therefore obtain our Huffman prefix code as

$$q = 0000, z = 0001, y = 001, o = 01, u = 10, a = 11$$

The code is as efficient as possible for the given frequency distribution. (However, there may be other equally good solutions—in particular we could swap 0 and 1 throughout if we wished.) The tree structure ensures that no code word will be a prefix of any other.

The number of 0's and 1's in the coded passage is the sum of the lengths of each code word multiplied by the number of times it appears in the text. In this case this gives us the sum of contributions from the letters q through to a in the order above as

$$(4 \times 4) + (4 \times 7) + (3 \times 12) + (2 \times 20) + (2 \times 28) = 176$$

Therefore the passage could be transmitted as a binary string of total length 176 and this is the shortest length possible.

Since we are dealing with a prefix code, the deciphering of the passage back to plain text will not require spaces or any other indicators. We would simply begin at the left-hand end of the coded message and read the binary string until the enciphered form of one of the six letters was recognized. Since this is a prefix code, the deciphering is unambiguous—there is no possibility that this string is merely an initial segment of a longer string of another code word. Having properly deciphered the first letter, we continue with the remainder of the code string in this fashion, detecting one letter at a time, until the entire message is revealed.

Reassembling RNA chains

RNA, like its better known cousin DNA, is genetic material carried within living cells. Ribonucleic acid, to give its full title, forms chains made up of four bases: uracil (U), cytosine (C), adenine (A), and guanine (G). RNA chains, however, break up into fragments in the presence of enzymes.

This happens in one of two ways. A G-enzyme breaks a chain after each G-link, whereas a U, C-enzyme breaks a chain after each U- and each C-link. For example, if the initial RNA chain was

GUGAUGACCAGCC

then the G-enzyme would act to give the fragments

G-fragments: *G, UG, AUG, ACCAG, CC*

while the effect of the U, C-enzyme would result in the set of fragments

$$U, C\text{-fragments: } GU, GAU, GAC, C, AGC, C$$

The problem encountered is that the research worker may be left only with the fragments, appearing in any order, and needs to reconstruct the original chain from the evidence to hand. The problem is known as that of reconstituting the RNA chain from its *complete enzyme digest*. It turns out that all possible solutions (sometimes there are more than one) can be reconstructed through finding all directed Eulerian paths in a certain digraph built in a way that will now be described, using this problem as a case study. We assume that we do not possess the original chain and, in any instance, we will be interested to find all other solutions, if there are any, other than the chain which is the seed of our problem. We also take it for granted that there are at least two fragments of both the G- and U, C-kind, for otherwise the solution is apparent.

Before arriving at the magic digraph, there are some elementary observations that make life easier.

It is possible to spot at once the fragment that forms the (right-hand) end of the chain. In the above example, the G-fragment CC does not end in G and is so-called *abnormal*, for the only way a G-fragment may end in a base other than a G is if it appears at the end of the chain. It is possible, although not a feature of this example, to witness an abnormal U, C-fragment—one that does not end in a U or a C. Indeed it is sometimes possible for both fragment lists to have one abnormal fragment (necessarily ending in A). In this case the abnormal fragments are different and both form part of the end of the chain so that the longer of the two will be a continuation of the shorter.

In summary, there is at most one abnormal fragment in each list and each occurs at the right-hand end of the chain, the longer being an extension of the shorter. In our example we have that the reconstituted chain ends with the abnormal fragment CC. The other four G-fragments could still be arranged in $4 \times 3 \times 2 \times 1 = 24$ ways, so the problem yet offers many potential solutions.

To emphasize that the original order of the fragments is lost let us suppose they come to us in the following order:

G-fragments *AUG, UG, G, ACCAG, CC*;
U,C-fragments *C, GAU, AGC, C, GU, GAC*

We shall refer to this list as the *complete enzyme digest* (CED). The next stage is to list what happens if we were to split the given G-fragments and U, C-fragments further using the other enzyme. For instance, the G-fragment *ACCAG* would break down further into *AC, C, AG,* in the presence of the U, C-enzyme while the U, C-fragment *AGC* would be broken into *AG, C* in the presence of the G-enzyme. The resulting minor fragments are known as *extended bases* and those extended bases that are neither first nor last in one of the original fragments are described as *interior*. For instance, when the fragment *ACCAG* breaks down giving the extended bases as listed above, *C* is an interior extended base. Plainly, interior extended bases only arise when a fragment breaks into three or more extended bases.

A list must now be formed of all the interior extended bases that would arise when this secondary round of splitting was carried out. In this case there is but one: *C*.

We also need to list all the *unsplittable fragments*, that is those fragments on either the G- or U, C-list that do not split further under the action of the enzymes. In our case this list consists of

G, C, C

There are two unsplittable fragments that are not interior extended bases, namely *G* and one of the *C*. This must always be the case: the chain must begin and end with fragments that could not be split further and any other fragment that does not split under the action of either enzyme must arise as an interior extended base. Since we know that our chain ends with *CC*, it must therefore begin with *G*.

From the complete enzyme digest the required digraph is constructed as follows. Consider any fragment such as *AUG* in the CED that is not an extended base, that is to say it splits under the action of another enzyme. Draw two nodes labelled *AU* and *G* for the first and last extended bases in this G-fragment and join them by an arc,

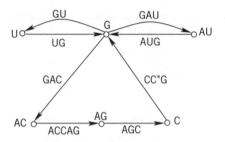

Figure 9.9 Digraph for reconstructing RNA

from *AU* to *G* in this case, carrying the label *AUG*. By the same token *UG* gives rise to an arc from a node labelled *U* to the node labelled *G* with the arc itself carrying the label of the fragment, *UG*.

We apply this procedure to all *normal* fragments that split. The result is represented in the digraph as shown in Figure 9.9. The digraph has one additional labelled edge not specified by the basic rule of construction. This final arc runs from the first extended base of the (longest) abnormal fragment, to the node labelled by the first extended base in the chain. In this example the abnormal chain is *CC* so that *C* is its first extended base, while the initial fragment of the chain is *G*, giving an arc directed from *C* to *G*, as shown. The label of this arc is CC*G.

The solution to the reconstruction problem now comes through reading the paths of Eulerian circuits that begin at node *G*, the initial fragment of the RNA chain, and end the circuit by returning to *G* along the special arc labelled in this case CC*G.

The actual solution is then given by the labels of the successive arcs, noting that each node is met *twice* on the arc label but should only feature *once* in the solution. For instance, in this example, we have an Eulerian circuit, the arcs labels of which yield the sequence

$$GU\,UG\,GAU\,AUG\,GAC\,ACCAG\,AGC\,CC^{*}G$$

which corresponds to the reconstructed RNA chain

$$GUGAUGACCAGCC$$

with which we began the problem. However, there is another Eulerian path that allows a different interpretation of the complete enzyme digest that has arc label sequence

$$GAU\,AUG\,GU\,UG\,GAC\,ACCAG\,AGC\,CC^*G$$

which gives the alternative solution

$$GAUGUGACCAGCC$$

The reader can easily check that this truly is a solution as the action of the G- and U, C-enzymes on this chain produces the same CED as given originally.

Another example that readers might like to try their hand on is the following complete enzyme digest:

G-fragments: $AACUG, UAG, A, AG, AG, AG, G$

U, C-fragments: $U, AGAAC, AGAGA, GGAGU$

On this occasion, the reconstructed chain is unique and the digraph features a loop carrying the starred label. The solution is given in the final chapter.*

The Eulerian circuits of the digraph, ending with the starred arc, represent all possible ways of consistently reconstructing the original RNA chain as the digraph carries all the information available from the complete enzyme digest, which is the starting point of the the problem. At least one solution to the problem must emerge as otherwise the given CED could not have come about in the first place. If by some chance the digraph failed to have an Eulerian circuit of the required type, that would indicate that some fault had occurred with the procedure, as that would not be possible otherwise.

10
For Connoisseurs

This final chapter is intended to give a little more mathematical explanation for those readers who would appreciate some without having to pursue another source. The level of difficulty will vary—much of it is not very hard at all but, unlike in the main part of the book, I will assume that the reader has some familiarity with mathematical ideas and notation. However, most readers will be able to gain from dipping into the text here and there.

In some places I will be using proof by induction. This is the mathematical technique where the proof is established by building from one case to the next. A case study that is relevant here is the fact that any tree with n nodes has $n - 1$ edges. We begin by checking the first case: if a tree has $n = 1$ node only, then it evidently has $n - 1 = 0$ edges. This *base case* anchors the induction. Next comes the general *inductive case* whereby we somehow show that *if* the statement held for trees with k nodes, then, in consequence, it will hold for trees with $k + 1$ nodes. Having established both the base case and the inductive step, it follows that the proposition holds for all trees. This is perhaps the most fundamental technique of mathematical proof, often likened to the toppling of an unending sequence of dominoes.

Chapter 1

Page 5 **Characterizations of trees**

Before we get into this, we make a simple observation that comes up time and again.

A network in which every node has degree at least two has a cycle.

To see this, start at any node in the network and set out to walk a trail, that is to say walk about the network, never retracing the same edge, in either direction, once you have passed over it. If, contary to our claim, we can never find a cycle in our walk, this will mean that each node we reach will not have been met before. Upon arrival at a node *u*, there will be at least one edge by which we may exit, as the degree of each node is at least two and *u* has not been visited before. Therefore if we never found a cycle, we could continue indefinitely without repetition of a node. This is, however, impossible as any network is finite (at least the ones we have considered in this book) and so we must eventually revisit a node, and in so doing trace out a cycle in the network.

   We *define* a tree as a connected network *N* that is free of non-trivial cycles. (A trivial cycle has length 0 and consists of starting at a node and not moving at all.) Suppose that our network has *n* nodes and *e* edges. The following are equivalent ways of defining the same idea, although this list is not exhaustive.

1. *N* is a tree;
2. There is a unique path between any two given nodes in *N*;
3. *N* is connected and $e = n - 1$.

In theorems of this kind there is no need to prove that each condition implies every other, which would in this instance call for some six separate arguments. The standard trick is to verify the implications cyclically. In this case we show that each of the three implications $2 \Rightarrow 1 \Rightarrow 3 \Rightarrow 2$ holds. Given this, we may begin at any of the numbered conditions and deduce any of the others as a consequence: for instance we may deduce 2 from 1 via 3.

2) implies 1). Given that condition 2) holds in N then there is certainly a path between any two nodes in N and so N is connected. Moreover, N cannot have a cycle of length more than 0 as a cycle clearly provides two distinct paths between any two points on the cycle. Therefore if 2) holds then so does 1).

1) implies 3). We proceed by induction on the number n of nodes of N, the base case where $n = 1$ being clear. Since N is connected, every node of N has degree at least 1. Given that N has no cycles, then N must have at least one *endpoint*, which is a node of degree 1, as this follows by the italicized observation at the beginning of this section. We remove an endpoint and its edge from N to give a new network N' with one fewer nodes than N. What is more, the network N' is still connected and is cycle free. Hence N' is a tree and by induction the number of edges of N' is $(n - 1) - 1 = n - 2$, so that our original network N has $n - 1$ edges.

3) implies 2). First we show by induction that a connected network of n nodes has at least $n - 1$ edges, a claim that is vacuously true if $n = 1$. Suppose then that a connected newtork N has $n \geq 2$ nodes. Delete as many edges as possible without disconnecting the remaining network N'. Then N' has no cycles (as otherwise it could be further pruned) and so has an endpoint u. Remove u and the edge incident with u from N' to give a connected network N'' on $n - 1$ nodes. By induction, it follows that N'' has at least $n - 2$ edges and so N', and hence the original N, has at least $(n - 2) + 1 = n - 1$ edges.

Now suppose that N is connected, that $e = n - 1$, but there are two distinct paths between nodes u and v in N. The two paths diverge at some point only to meet up again at some later point and in so doing form a cycle containing the two points where the paths first diverge and next meet. Removing an edge from this cycle would leave a network on n nodes that was still connected but had only $n - 2$ edges, contrary to what we just proved. Hence the path between any two nodes of N is unique, thus completing the proof.

Other similar characterizations of trees are as the connected networks with fewer edges than nodes; networks with no circuits and exactly one more node than edges; connected networks in which every edge is a bridge; and networks that are free of circuits but where the addition of any new edge between existing nodes creates one.

Another way to see that trees have $n-1$ edges comes through regarding the network as directed. Choose any node of your tree as its *root* and direct all edges away from the root; since there are no cycles this can be done unambiguously. Every node, except the root, has an edge associated with it, that being the final edge from the directed path from the root to the node. This gives a one-to-one correspondence between the edges and the non-root nodes so, in particular, the number of edges must be one less than the number of nodes.

Page 15 **Puzzling Liars**

Who broke the window? If Alex had done it then Barbara would not have blamed anyone else. Therefore, Alex is telling the truth and Barbara must be the culprit. This is also consistent with what Caroline and David said.

A says B is a liar and B says the two are from different tribes. If A were telling the truth then A would be a 'Knight' and B a 'Knave'. However, B's statement would then be true, contradicting the conclusion that B is a liar. Hence A is a liar, so B must be truthful and B's observation is both true and consistent. Therefore A is a liar and B is not.

Finally we come to the problem about the string of natives, each calling the next a liar and the final one claiming they are all liars bar him. Essentially the Knaves and Knights must alternate in this situation, but whether or not A is Knight or Knave depends on whether we have an even or odd number of natives in the line. Let us suppose that the total number of natives is even. If there were only two of them, we would have a pair of natives branding one another liars. One of them would be a Knave and the other a Knight but we would have no way of telling which. Suppose that the number of natives were 4 or 6 or some other even number. If we suppose that A is truthful, it will follow that B is not, that C is, and so on with all the odd-numbered natives truthful but the even-numbered ones are liars, and that goes for the last man as well. Since he branded all the rest liars and not all of them are, this is all consistent. However, if we assume that A is lying we get the same alternating pattern of Knaves and Knights but now it is the even-numbered natives, B,D,F, and so on that are the Knights. However, that would make the last man a Knight, but he claims that all of them, including the truthful B for instance, is a liar. This is inconsistent and so we conclude that if there is an even number of natives, then the first, third, fifth, and so on are Knights but the rest are lying Knaves.

A similar argument applies if we are faced with an odd number of natives but now, as you can easily check, the opposite conclusion applies: A is a liar and now it is A,C,E, and so on who are the Knaves, which includes the last man, and the rest are Knights.

Chapter 2

Page 23 Number of weighings for the counterfeit coin

In general, the number of weighings to solve this problem is $\log_3 n$, rounded up, where n is the number of coins. We proceed just as in the particular problem given in the text by weighing in order to divide the current subset of coins to hand into three almost equal piles by comparing two piles of the same size. We can always do this whether or not the total number of coins remaining is or is not a multiple of 3. The outcome tells us each time which of the three piles contains the fake. The number of weighings needed will be the *height* h of the ternary tree that starts with a root and each node, except the endpoints, known as the *leaves*, has three offspring nodes. The value of h will be the least number where the final generation has at least n nodes, covering all the possibilities for the identity of the fake.

In general, the number of nodes of each new generation of an m-ary tree (one where each internal node has m offspring) is m times the previous level so that the number of nodes at each level follows that pattern $1, m, m^2, \ldots, m^h$. In our problem we have a ternary tree so that $m = 3$ and the value of h is therefore the least integer such that $3^h \geq n$. Taking logarithms to the base 3 then gives $h \geq \log_3 n$, and since h must be a whole number, the value of the logarithm should always be rounded up to the nearest integer to give the minimum number of weighings. In the example in the text, $n = 9$ and so $h = \log_3 9 = \log_3(3^2) = 2\log_3 3 = 2$, and, as we saw, two weighings sufficed.

Page 25 Numbers of Latin Squares

We need the exclamation mark *factorial* notation here: $n! = n \times (n-1) \times (n-2) \times \ldots \times 2 \times 1$. The number of Latin squares of a given size n is huge: at least $n!(n-1)!(n-2)!\ldots 2!$ This is because the first row can be filled in $n!$ ways, then there are $n-1$ choices for the first entry of the second row, at least $n-2$ possibilities for the second entry, and so it continues. This is only a lower bound as, for example, there *may* be $n-1$ choices for the second entry of the second row in the event that entries in positions $(1,2)$ and $(2,1)$ are identical. For $n = 3$ there are exactly $3! \times 2! = 6 \times 2 = 12$ Latin squares. For $n = 4$ there are however 576 Latin squares although $4! \times 3! \times 2! = 24 \times 6 \times 2$ is only 288. Even the number of Graeco-Latin squares, also known as *orthogonal Latin squares*, grows fast: the numbers of orders 3 and 4 are respectively 36 and 3456.[1]

It should be said, however, that many of the Latin squares are equivalent in that one can be transformed into the other simply by permuting, that is to say renaming, the entries or through one of the eight symmetries of the square (reflections in sides or diagonals; rotations through one or more right angles about the centre). Two Latin squares related in this way might be regarded as essentially the same. Indeed when discussing Latin squares it is often taken as read that they are *normalized*, which means that both the first row and column consist of the numbers $1,2,\ldots,n$ in that order, for there is no loss of generality in doing this for most matters of interest.

---

[1] This and other mathematical facts can be called up quickly from the excellent website <http://mathworld.wolram.com>, which links and leads to original sources.

Page 27 **Difference of Two Squares**

Any odd number $2m+1$ and any multiple $4m$ of four is the difference of two squares by virtue of the identities $2m+1 = (m+1)^2 - m^2$ and $4m = (m+1)^2 - (m-1)^2$. However, for any difference n of two squares we have $n = a^2 - b^2 = (a-b)(a+b)$ and since the factors on the right differ by the even number $2b$, they are either both even, in which case n is a multiple of 4, or these factors are both odd, in which case so is n. Therefore the numbers of the form $4m+2$ are exactly the ones that *cannot* be expressed as the difference of two squares. (See Figure 2.6 on p. 29.)

3	2	6	7	8	4	1	9	5
7	8	5	6	1	9	2	4	3
1	9	4	2	5	3	8	6	7
5	1	7	4	6	8	9	3	2
4	3	2	5	9	1	6	7	8
8	6	9	3	2	7	5	1	4
9	5	3	8	7	6	4	2	1
6	7	8	1	4	2	3	5	9
2	4	1	9	3	5	7	8	6

Figure 10.1 Sudoku puzzle solution

Pages 28–30 **Sudoku Puzzle Solution**

For the Circular Sudoku the solution is displayed in Figure 10.2 (see Figure 2.8 on p. 30).

The circular version may look tricky because of the condition on overlapping quarter circles. It would seem that you have to keep track of a series of overlapping sets and that looks like a headache. It is this that makes the problem easy, however. If you look at any slice of the pie making up $\frac{1}{8}$ of the circle, the slice will contain four numbers. The slices either side of it must both contain the complementary set of four numbers, and this applies throughout the circle. In other words, if you colour the slices alternately red

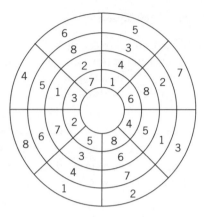

Figure 10.2 Circular Sudoku solution

and blue, the red slices contain the same set of four numbers, and the blue slices similarly carry the complemetary four. The first step, then, is to collect up the 'blue' and the 'red' numbers. Let us say that the segment at the top right headed by the number 6 is a blue slice. Gathering up the numbers from the other blue slices then gives us the blue set $B = \{6,2,8,7\} = \{2,6,7,8\}$, and so in this case the red set is $\{1,3,4,5\}$. The 4×4 array making up the blue slices is then a 4×4 Latin square in the numbers from B while the reds similarly form a 4×4 Latin square in the red numbers. There are enough numbers in each to determine each of these squares completely and finding the solution is now child's play.

Armed with this advice, you might like to solve the next example that is one size up. This time, each of the ten numerals 0 through to 9 must appear in each of the five rings and each of the ten double slices. Again, the solution is unique (see Figure 10.3).

There are other variants of traditional Sudoku although they generally seem to involve a bigger and more complicated version of the same array. This new puzzle takes a leaf out of the book of Graeco-Latin squares in that it involves two parallel Latin squares harnessed in tandem. The idea, however, is not to *superimpose* the squares but rather to *interleave* the rows of the pair and wrap them around, bottom to top. The effect is the circular array shown. The charm

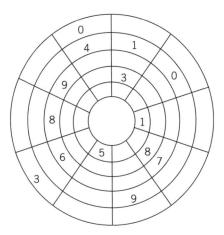

Figure 10.3 Five-ring Circular Sudoku

of the puzzle lies in the way the Latin squares smuggled themselves into the picture and split it into two parallel problems.

What is more, only $\frac{10}{32}$, which is about 31 per cent, of the numbers needed to be given in order to determine a unique solution, which is somewhat less than some orthodox Sudoku. There are examples, however, of Sudoku puzzles with as few as 17 'givens'. At the time of writing it has not yet been established whether or not 17 is the minimum number possible. In the given square Sudoku problem the proportion of filled cells at the start was $\frac{30}{81}$, which is about 37 per cent. However, the least number of cells that can fix the solution is 9 in the 4-ring version and, I believe, 13 in the 5-ring puzzle, which repesents only 26 per cent of the entries.

The original format as appeared in newspapers in 2005 lacked the space in the centre of the puzzle. In *The Official Book of Circular Sudoku* we opted to punch a hole out in the middle of the circle so that each ring, include that innermost, appears as a true ring. The purpose of this is just to avoid the cramping of symbols near the centre of the diagram and does not alter the mathematical nature of the puzzle in any way. Also light shading of every other ring makes it easier for the eye to follow a ring around the puzzle without losing track.

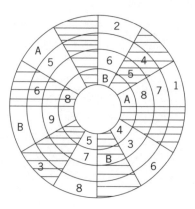

Figure 10.4 Target Sudoku Puzzle

A more diabolical variation involving only partial overlap conditions can be manufactured. These do not split into two distinct parts and pose dilemmas of the kind that arise in the tougher standard Sudoku challenges. In the puzzle of Figure 10.4, the challenge is to fill each of the four rings with each of the twelve symbols $0,1,2,\ldots 9,A,B$ in such a way that each of the *six* white-black-white quarter circles also features every one of the twelve symbols. There is, however, no rule concerning the black-white-black quarter circles. If the same rule did apply, the problem would split into three parallel 4×4 Latin squares and would be easily solved. As it is however, with fewer constraints on the array, it takes more work to find the unique solution (Figure 10.5).

Page 40 **Power Laws**

To say that one quantity y is related to another x by a *power law* means that there is a relationship between them expressed by a rule of the form $y = kx^n$, where k and n are fixed numbers. We say that *y is proportional to x* to the power n. In the context of nodal degrees in a network, y would represent the number of nodes of degree x and so as x increases we would expect y to fall off quite sharply. For this to occur, the exponent n needs to be negative or, put differently, the law has the form $y = \frac{k}{x^m}$ where m is some positive number. As x

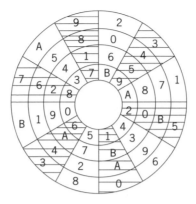

Figure 10.5 Solution to Target Sudoku

increases, y decreases in these circumstances and the larger the value of m, the more rapid the descent. This is the kind of power law that seems to arise in many real networks. The value of m itself has to be estimated and may take on a value other than a whole number. However, in some important examples the value of m seems to be about 3.

In the case of random networks, the distribution of degrees of nodes does not follow a power law but rather we see what is known as *exponential decay* characterized by laws like $y = \frac{1}{2^x}$. In this kind of law the base of the exponent is not x but rather a fixed number, in this case 2, but other values arise. The exponent, however, is *not* a fixed number n but is x itself. This makes an enormous qualititative difference—an exponential decay is, in the long run, much more rapid and severe than one determined by a power law.

It is true that for any large value of x, both types of law will return very small values of y and the larger x the smaller the y. However, if we take the value of y as given by a power law and divide it by the y-value as provided through an exponential law we find that, for large values of x, the ratio is very large and keeps getting larger. In other words, although both quantites are small, the power law value will be many thousands or even millions of times larger than that provided through exponential decay which vanishes with extraordinary rapidity.

This accounts for the hubs. In a 'power law' network, we will see some large hubs and a few very large ones. In a random network we are more than likely to see no large hubs at all, the largest nodal degree only being two or three times the size of the average.

In the literature, networks subject to these power law distributions are sometimes described as *scale-free*. This is *not* intended to mean that their structure has a fractal-like property of looking the same when examined on any scale, whether it be very large or at the finest level. Rather, it is an implicit reference to another qualitative difference between a random distribution and one based on a power law.

In the case of a random distribution, the graph of numbers of nodes versus the degree of the node displays a typical bell-shaped or *normal* curve that is nearly symmetric and is dominated by a well-defined peak. The degree corresponding to this peak then gives the measure of the degree of a typical node and in that way lends a scale to the entire network. This contrasts with a power law distribution which falls away continually as we move to the right with a larger degree value always being less likely than a smaller one, despite the fact that we see more examples of large hubs in networks subject to power laws than we observe in random networks. These networks are *scale free* in the sense that even though it is possible to calculate the mean size of the degree of the nodes in the network, that average does not represent a particularly significant statistic. For example, even though the average degree, say, was six, there would be more nodes of degree five than degree six.

The term *small world networks* has a mathematical meaning, which is that of a large random network where most nodes are not mutual neighbours yet the average path length between randomly chosen nodes is small. Small world networks do not necessarily have the scale free property even though this is commonly the case in large social and other naturally occurring networks in biology and physics.

There is a simple piece of mathematical trickery for detecting a power law. Suppose we suspect that y is related to x by a power law but we have no idea what the values of k and n might be. We can still flush the law out by use of the venerable mathematical device of logarithms, invented by the Scot John Napier around the turn of the seventeenth century.

If a law of the form $y = kx^n$ is present then, by taking logarithms to any base we obtain

$$\log y = \log(kx^n) = \log k + \log x^n = \log k + n \log x$$

The last two equalities are justified through use of the so-called log laws: the log of a product is the sum of the logs, and the log of a power is the exponent times the log. Since $\log k = A$, say, is a constant, as is n, it follows that if a power law is present then we will observe a tell-tale linear relationship when we plot $\log y$ against $\log x$ on graph paper. Moreover, if this linear relationship emerges, so will the values of the unknown constants k and n, for n will be the gradient of the line and the value of $A = \log k$ will equal the coordinate of the intercept of the vertical axis.

Traditionally, this technique was called on so often in the physical and biological sciences that log-log paper was invented—graph paper where the axes were already scaled logarithmically so that plotting data subject to a power law will immediately manifest itself in a straight line graph on the log-log paper.

Chapter 3

Page 59 **Existence of Ramsey Numbers**

We will show you how to find bigger cliques at bigger parties. The next question along in the sequence is: How large a party do we have to have in order to ensure that there is a group of four mutual friends or four mutual strangers? Cast in the language of networks, we ask: How many nodes does a simple network N require in order to ensure that either N or its complement N' contains a copy of K_4, the complete network on four vertices? It has been verified that the answer is 18. I cannot show that here. What I can prove, however, is that the number does exist, and that it is no more than 63.

This may not sound too impressive, but remember that it is not obvious that the number has to exist at all. What is more, although we will run through the argument for 4-cliques, it does extend in a very straightforward way to show the existence of all Ramsey numbers and indeed the argument for $m = 4$ is completely representative of what happens in general. In contrast, the simple argument in the text that works for K_3 does not generalize to higher Ramsey numbers without some extra complications arising.

The argument demonstrates that given any m, there is a number n such that if a network N has at least n vertices then either N or its complement contains a copy of the complete network K_m, and what is more the argument shows that n need be no more than $4^m - 1$. This bound may vastly exceed the true size of the Ramsey Number in question but it does show copies of K_m are inevitable in sufficiently large network-complement pairs.

The argument even has useful interpretations in cases involving infinite sets and the whole flavour is that of the Pigeonhole Principle, which you will see popping up explicitly throughout the proof.

It is best to imagine the network N and its complement N' to be superimposed, giving a copy of the complete network on n vertices but, in order to keep track of which edge belongs to which network, let us colour, in our imaginations at least, the edges of N blue, and the edges of N' in red. What will be shown is that, provided that N has at least 63 nodes, this complete network must contain a *monochromatic* copy of K_4; that is to say, there is some set of four nodes with the edges running between them all coloured blue, or all red.

Suppose then that our network (or party, if you prefer) has at least

$$1 + 2 + 2^2 + 2^3 + 2^4 + 2^5 = 63 \text{ nodes.}$$

The precise value of this number is chosen only to ensure that we have a sufficiently large supply of nodes to carry out the following procedure without running out.

Focus on one node—A_1 say—and proceed as follows (see Figure 10.6). Of all the edges leading from A_1 (there are at least 62 of them, of course), at least half will be of one particular colour, let us call that colour C_1 (C_1 will either be blue or red). Consider all the

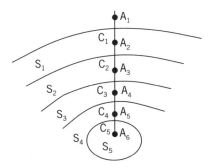

Figure 10.6 The inevitable four-clique

nodes connected to A_1 by an edge of colour C_1, and call this set of nodes S_1. There are at least

$$\frac{1}{2}(2 + 2^2 + 2^3 + 2^4 + 2^5) = 1 + 2 + 2^2 + 2^3 + 2^4 = 31$$

of these nodes. (Of course it is obvious that half of 62 is 31, the calculation is displayed this way here only to draw attention to the general pattern that emerges.) Let us choose one of them and name it A_2.

At least half the nodes from A_2 *leading to other nodes in* S_1 are of the one colour; call this colour C_2, which may or may not be the same as C_1. Let S_2 denote the collection of these nodes. Note, and this is critical, S_2 is entirely contained in S_1 as indicated in the diagram and S_2 itself has at least

$$\frac{1}{2}(2 + 2^2 + 2^3 + 2^4) = 1 + 2 + 2^2 + 2^3 = 15$$

members. Choose a member of S_2, calling it A_3.

We carry out this process five times in all, giving us nodes $A_1, A_2, \ldots, A_6$ and a descending chain of sets

$$S_1, S_2, S_3, S_4, S_5$$

each of which is contained in the one before, as indicated in Figure 10.6. It is now clearer that the purpose of the choice of the initial number of nodes (63) was to guarantee that we can carry out this process at least five times—the sets S_3, S_4, and S_5 will have at least 7, 3, and 1 member respectively.

How does all this help? We need one subtle observation now to settle the question. The next paragraph has the key idea, although it requires a little thought.

Consider the list of nodes A_1, A_2, A_3, A_4, A_5. Look at any member in the list, A_3 say. All the edges from A_3 to the members of the set S_3 are of the one colour. Now the nodes A_4, A_5, and A_6 *are all in* S_3 so that all of the edges from A_3 to the members of the list that follow A_3 are the same colour. This argument applies equally well to all the nodes A_1 through to A_5: each of the A_i's has a colour associated with it, C_i, the colour of the edges leading from it to all the members of the list that follow it. Now there are only the two colours available, blue and red, and so, by the Pigeonhole Principle, at least three of $A_1, A_2, \ldots, A_5$ have the same colour (blue say) associated with them. Choose such a group of three nodes together with A_6; now *every* edge between these four nodes must be blue, and so we have discovered our required monochromatic copy of K_4 or, if you prefer, we have tracked down a clique of four mutual acquaintances at the party of sixty-three or more people.

Ramsey numbers are normally defined slightly more generally. A set of nodes in a network is called *independent* if none is adjacent to any other in the set; this corresponds to saying that the set form a clique, that is a complete graph, in the complementary network. The Ramsey number $R(k,l)$ is then defined to be the least number n such that every network with at least n nodes contains a copy of the complete network on k nodes *or* an independent set of size l. If $k = l$ we get the Ramsey numbers that we have been talking of thus far. The argument of Chapter 2 shows us that $R(3,3) = 6$ and the Ramsey numbers for smaller values of k and l are easily found: indeed $R(k,2) = k$ for all values of k. There are some simple observations: the values of $R(k,l)$ increase monotonically in each of the variables k and l and the function R is symmetric in that $R(k,l) = R(l,k)$ because a network N satisfies the requirements of the $R(k,l)$ condition if and only if its complementary network N' satisfies the requirements of the condition defining $R(l,k)$.

The above argument shows that $R(4,4) \leq 63$. In fact it is known that $R(4,4) = 18$ and also $R(3,4) = 9$ (not hard to show); moreover $R(3,5) = 14$, $R(3,6) = 18$. Importantly, as already mentioned, the argument type extends naturally to show that all the Ramsey numbers $R(k,k)$ exist from which it follows that $R(k,l)$ always exists as this number is certainly bounded by $R(m,m)$, where m is the maximum of the two numbers k and l. However, exact values of the function R are not known for many other pairs: for instance $R(4,4)$ may be as small as 43 but the question is undecided. However, there is a useful bound in terms of binomial coefficients:

$$R(k, l) \leq C(k + l - 2, k - 1) = \frac{(k + l - 2)!}{(l - 1)!(k - 1)!}$$

Applying this inequality gives quite sharp upper bounds for small Ramsey numbers: $R(3,3) \leq 6$, (the exact value) $R(4,4) \leq 20$ (compared with the exact 18), but the inequality only tells us that $R(5,5) \leq 70$, when the exact answer is known to lie in the range 43–49.

This is a glimpse of the tip of the iceberg that has become Ramsey Theory, which centres on results of the kind that show that in a sufficiently large 'system' there are subsystems of a given size with more organization and structure than the original object was assumed to possess. Another classic result of this genre that has spurred a whole theory in itself is Van der Waerden's Theorem that says that if you partition the positive integers into two classes in any way at all, then at least one of the classes contains arithmetic progressions of arbitrarily long lengths.

Page 61 **Pigeonhole Principle**

The claim is that any set of $n + 1$ numbers from among the first $2n$ positive integers must contain one number that is a factor of one of the others. First we observe that any number m can be written in the form $m = 2^k t$, where $k \geq 0$ and t is odd. The index k will be zero exactly if m is already odd and the number t will be 1 if m happens to be a power of 2. Given that m lies in the range from 1 to $2n$, so does its odd factor t. However, there are only n distinct odd numbers in this range so that it follows, by the Pigeonhole Principle, that two different numbers from our subset of size $n + 1$ share the same odd factor t. Call these numbers m_1 and m_2 so that we have $m_1 = 2^{k_1} t$ and

$m_2 = 2^{k_2}t$, say. The smaller of these two numbers, m_1 and m_2, is then a factor of the other, as required.

Once again, this inference is tight—if we replace $n + 1$ by n the claim is false: we only need consider the set of n numbers $n + 1, n + 2, \ldots, 2n$ to see this.

If you are up for another challenge, try to show, using the Principle, that given any eight numbers, the sum or difference of at least one pair of them must be a multiple of 13, although this is not necessarily the case if you begin with only seven integers.

Chapter 4

Page 77 **The five-colour map problem**

Enough has been revealed in the text to give a demonstration showing that no more than five colours are ever required to colour a map, however complicated. Since we can work one component at a time, we need only consider a typical connected planar network, N. We show, by induction on n, the number of nodes of N, that the network may be 5-coloured. There is obviously no trouble anchoring the induction as a one-vertex network can be 1-coloured.

We make use of the fact, explained in the text, that any planar network must have a node x of degree no more than five. We assume inductively of course that any planar network on fewer than n nodes has a chromatic number no more than five. If we remove x from N, along with its edges we have a planar network on $n - 1$ nodes which can, by induction, be 5-coloured (whether or not is is connected is irrelevant as that is not part of the inductive assumption). We now replace x to recover the network N and the task is to find a way of properly completing the colouring with x included.

We next clear the decks by disposing of the easy cases. If the degree of x is actually less than five, we have no problem, as we can simply colour x differently from any of its neighbours and the colouring is complete. The same applies even if the degree of x is five if two of its neighbours share the same colour—we colour x with a colour that has not been used by any of the neighbouring vertices.

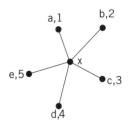

Figure 10.7 Five-colouring a network

The hard case then is where x has five neighbours, one of each of the five colours we are using. This crucial case is depicted in Figure 10.7 where we have numbered x's neighbours 1 through to 5 in clockwise order and used the letters a,b,c,d, and e to stand for the colours currently assigned to each of these nodes.

First we look at all paths from a whose nodes are coloured alternately 1 and 3. We can change the colour of a from 1 to 3, change the neighbours of a coloured 3 to 1, and so on along all paths of 1- and 3-coloured nodes coming from a and this will not violate the colour condition, as all nodes adjacent to these and not involved in the change of colour necessarily carry colours other than 1 and 3.

Suppose there is no path of this kind that reaches from a to c. This interchange of the 1 and 3 colours then cannot affect the colour of c and since a and c are now both coloured 3, we can safely colour x with a 1.

The alternative is that there *is* a 1–3 path from a to c. In this case consider all paths alternately coloured 2 and 4 beginning at b. Then b and d lie on opposite sides of a 1–3 cycle that forms a 1–3 path from a to c and the edges (c,x) and (x,a) and so there is no 2–4 path from b to d, as where two paths cross, the common node has to be coloured consistently with both paths. We can therefore perform a 2–4 switch along all 2–4 paths from b leaving nodes b and d both coloured 4 and allowing us to complete the colouring by painting x the colour 2.

This completes the inductive step and so we have proved that any map can be five-coloured.

It is tempting to try and work this inductive argument with four colours as the node e seems to have played little role. However, of the five neighbours of x, two will now carry the same colour and this forces the argument to run into difficulties that, it seems, cannot be circumvented at all easily.

Page 81 **Characterizing 2-colourable networks**
(See entry for page 165 below.)

Page 83 **Guarding the Gallery**

The gap left in the argument was how to show that an n-gon can be triangulated with $n - 3$ non-crossing diagonals, a fact which is vacuously true for the base case when $n = 3$, as we already have a triangle, so let us assume that n is at least four. The idea is to split the polygon P into two smaller ones using *one* diagonal, from which point the inductive hypothesis takes charge and sees us through.

To locate a suitable diagonal we first observe that it is not possible for *all* of the corners of the museum to be reflex angles, that is angles exceeding $180°$; this is because there are n interior angles in P and the sum of these angles is, by an elementary geometric argument, $(n - 2)180°$. Hence there is a *convex* interior angle A inside P. (In fact there are at least three of them as if there were fewer than three, the degree sum would be exceeded.)

Now let us look to the two neighbouring vertices B and C of A. If the segment BC lies entirely inside P, then this can be our diagonal. If not (the pictured case), then the triangle ABC contains at least one other vertex (this conclusion requires that A is not a reflex angle). Slide BC directly towards A (see Figure 10.8) until it strikes its last vertex D in ABC. Now AD lies within P and we have our diagonal.

The argument can now be completed inductively by considering the two polyhedra P_1 and P_2 that have the common diagonal AD and whose other sides comprise all the sides of P. The number of vertices of P_1 and P_2 are respectively m_1 and m_2, say, where $m_1 + m_2 = n + 2$ (as P_1 and P_2 have vertices A and D in common). By induction we may triangulate P_1 and P_2 using no more than $m_1 - 3$ and $m_2 - 3$ non-crossing diagonals respectively. These two sets of diagonals together with the common diagonal AD yields a required

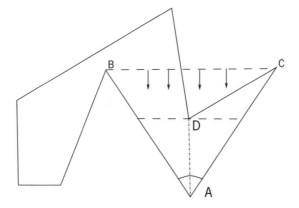

Figure 10.8 Locating a splitting diagonal

triangulation of P using no more than

$$(m_1 - 3) + (m_2 - 3) + 1 = (m_1 + m_2) - 5 = n + 2 - 5 = n - 3$$

diagonals in all, as required. This fills the gap left in our proof and allows us to be sure that we never need more than $\frac{n}{3}$ guards to mind the museum.

Page 99 Brouwer's Theorem

Recall that, in two dimensions, the theorem says that any continuous mapping of the closed disc into itself has a fixed point. The shape of the figure concerned is of no importance (but it does need to contain its boundary). Instead we work with triangle Δ with vertices $e_1 = (1,0,0)$, $e_2 = (0,1,0)$ and $e_3 = (0,0,1)$. This is enough, for the disc can be mapped in a one-to-one and continuous fashion onto Δ, whence it follows that a continuous self-mapping of the disc with no fixed point would yield a mapping on Δ with the same property. Therefore we need only show that any continuous mapping $a : \Delta \to \Delta$ taking Δ into itself leaves some point where it is.

Recall the meaning of a *triangulation T* of a figure: we partition the object into triangles in such a way that one side meets another only at common corners. By the *mesh* $\delta(T)$ of the triangulation we mean the length of the longest line segment that can be drawn within the figure without crossing a side of one of the triangles involved in T.

The idea is to construct an infinite chain of triangulations of Δ; $T_1, T_2, \ldots$ such that the sequence of meshes $\delta(T_k)$ converges to 0. This can be done, for example, by triangulating each little triangle from its centre of gravity, known as its *barycentre*, in order to get the next in the sequence of triangulations.

For each of these triangulations, we introduce a 3-colouring of the vertices v by setting $c(v) = \min\{i : a(v)_i < v_i\}$, that is $c(v)$ is the least index i such that the ith coordinate of $a(v) - v$ is negative. Assuming that a has no fixed point, this is well-defined for the alternative is that $a(v)_i \geq v_i$ for all i with strict inequality in at least one case; now every $v \in \Delta$ lies in the plane $x_1 + x_2 + x_3 = 1$, and hence $v_1 + v_2 + v_3 = 1 = a(v)_1 + a(v)_2 + a(v)_3$, and so, given that $a(v) \neq v$, at least one of the three coordinates of $a(v) - v$ must be negative, and at least one must also be positive.

The colouring that now arises respects the hypotheses of Sperner's Lemma. Each vertex e_i must be assigned the colour i since its ith component is the only one of the components of $a(e_i) - e_i$ that can be negative as the other two components of e_i are both zero. Moreover, if v lies on an edge opposite e_i, then $v_i = 0$ and so the ith component of $a(v) - v$ *cannot* be negative, and hence v is *not assigned* the colour i. The hypotheses of the lemma are therefore satisfied and so we look to see what follows from its conclusion.

Sperner's Lemma now assures us that in each triangulation T_k there is a triply coloured triangle $\{v^{k,1}, v^{k,2}, v^{k,3}\}$, with i being the colour of the vertex $v^{k,i}$. Now the sequence of points $(v^{k,1})_{k \geq 1}$ need not converge but the sequence must have a subsequence that does converge to a point v in Δ. (This is a *compactness* property: this is the point in the argument where we require that the set Δ is both bounded and *closed*, that is to say Δ contains its own boundary.)

We now replace the chain of triangulations T_k with the corresponding subsequence although, for simplicity, we continue to denote the members of this subchain by T_k. Now the distance of the points $v^{k,2}$ and $v^{k,3}$ from $v^{k,1}$ is at most the mesh,

$\delta(T_k)$, which converges to 0. It follows that the sequences of points $(v^{k,2})$ and $(v^{k,3})$ also both converge to one and the same point v.

The question now to ask is: What is $\alpha(v)$? We have designed the colouring so that the first coordinate of $\alpha(v^{k,1})$ is smaller than that of $v^{k,1}$ for all k. Since α is continuous and the sequence of $v^{k,1}$ converges, it follows that the first coordinate of $\alpha(v)$ is less than or equal to that of v. The same reasoning applies to the second and third coordinates, yielding the inference that none of the coordinates of $\alpha(v) - v$ is positive, a conclusion we have already seen contradicts the assumption that $\alpha(v) \neq v$. This therefore completes the demonstration of the Brouwer Fixed Point Theorem.

The compactness property features quite critically in the proof and is not just a technicality that might be circumvented by some other argument—if we begin with the unit disc stripped of its boundary, then the theorem does not apply. For example, we could consider the mapping of the disc that acts as follows. We map each point (x,y) in the open unit disc to a point moved right by half the horizontal distance to the circle's circumference: $(x + t_x, y)$, where

$$t_x = \frac{\sqrt{1 - y^2} - |x|}{2}.$$

This is a continuous mapping whose range is the whole open disc. However, it has no fixed points as for any point in the *open* disc, $t_x \neq 0$; indeed $t_x > 0$ as $|x| < \sqrt{1 - y^2}$ for any point of the open disc. We see, therefore, that inclusion of the boundary matters in the fixed point theorem.

Chapter 5

Page 104 **Euler circuits**

What remains to be proved is that a connected network N in which every node has even degree does possess an Euler circuit. We can prove this by induction on the number of edges of N. Since N is connected, every node is even and none are isolated so that the degree of every node is at least two. By the very first note proved in this chapter, this guarantees that N does have a circuit, C. If C contains every edge of N then we are finished; if not, delete the edges of C from N to give a new network G which may well be disconnected, but in which every node still has even degree as the degree of every node of N has decreased by a multiple of 2, if at all. Now each component H of G is connected and has nodes only of even degree so that, by induction, each such H has an Eulerian circuit. The idea now is to sew these circuits back onto C to create a grand circuit for N.

More precisely, we begin on C and traverse that circuit until we meet a node u that is common with one or more of the components H. Inductively we now trace an Euler circuit for each such H based at u. When this process is exhausted we continue to move on around C, repeating this procedure where necessary when we encounter new components, until we have returned to our starting point along C. Since every component H meets C at some point, the resulting grand circuit represents an Euler circuit of the original network N.

A similar inductive argument can now be used to show that the construction given in the Fleury algorithm can be carried out and that the construction always yields an Eulerian circuit.

The feasibility of the Fleury algorithm is not so obvious as it makes the demand that we can recognize when an edge of a network forms a bridge. This sounds innocent enough but in effect it requires us to be able to tell whether a given network is or is not connected. We can, however, solve this problem by building a maximal spanning tree within the network using either the Prim or Kruskal algorithm and these procedures have no more steps than the given network has edges. Given what we now know (by the above proof) that an Euler circuit does exist in a connected network with no odd nodes, it is not hard to convince yourself that the Fleury algorithm will find one for you: by design, after each step, the remaining network is connected and so the procedure will not stop until every edge has been traversed.

Chapter 6

Page 112 **Hamiltonian tournaments**

The claim is that every tournament T is semi-Hamiltonian, meaning there is a (directed) path that travels through every node just the once. This is clear for the cases where T has fewer than three nodes so we shall assume that T has at least that many vertices. Assume inductively that any tournament with n nodes has a Hamiltonian path and consider a tournament T with $n+1$ nodes. Take one node v and its arcs away to give an n-node tournament T' that we take inductively to have a Hamiltonian path $v_1 \to v_2 \to \ldots \to v_n$, say.

Now either there is some arc $v \to v_i$ in T or not. In the first case, let i be the least index where this is true. We then have the following Hamiltonian path in T:

$$v_1 \to v_2 \to \ldots v_{i-1} \to v \to v_i \to \ldots \to v_n$$

because there is an arc $v_{i-1} \to v$ as there is no arc $v \to v_{i-1}$ and T is a tournament; note that if $i = 1$ then the above path simply begins at v. In the alternative case, no such i exists, in which case $v_n \to v$ is an arc of T and this arc may be tacked on the end of the Hamiltonian path of T' to give the required Hamiltonian path for T. Therefore any tournament has a Hamiltonian path.

It can be shown by similar argument that if T is strongly connected, that is to say there is a directed path from any node to any other, then T has a Hamilton cycle. Indeed more can be proved: in these circumstances T is guaranteed to have directed cycles of all lengths $3, 4, \ldots n$ (see for example Robin Wilson, *Introduction to Graph Theory*).

However, in general a digraph that is not a tournament may be strongly connected but lack a Hamiltonian cycle. For example, take the bow-tie network of Figure 3.7 and direct its edges $A \to B \to C \to D \to E \to C \to A$. This directed circuit shows the digraph is strongly connected yet there is still no Hamilton *cycle*.

Page 123 Constructing some automata

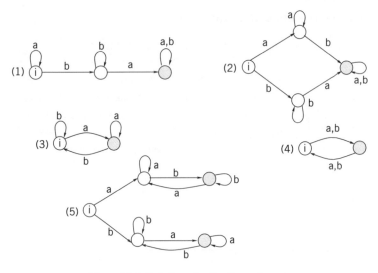

Figure 10.9 Solutions for the five automata

Minimal automata accepting the languages (1) through to (5) are seen in Figure 10.9.

Page 126 Algebraic realization of regular sets

This characterization is often not well appreciated even by experts in the subject of automata, at least if they have come to the topic from the direction of engineering. It furnishes a beautiful mathematical framework in which to operate. It does, however, involve algebraic semigroups when most mathematically trained people are only familiar with groups.

The story goes like this. A group is a set with an associative binary operation, the action of which can be reversed to return the identity element. In a semigroup, we drop the requirement that inverses need exist in this way and we do not even demand that a semigroup have an identity element, although this is a matter of less importance. A fundamental type of group is the group of all permutations of a set with the operation being function composition. A classic result of Cayley shows that any group can be realized as a group of permutations on a set. The counterpart in semigroup theory is the semigroup of all *functions* on a set. This semigroup is a *monoid* as it does possess an identity element in the usual identity function on the set. However, since functions are not in general one-to-one or onto, the members of this semigroup, known as the *full transformation semigroup*, do not in general have inverses in the sense needed in a group. Cayley's theorem for semigroups is that any semigroup can be realized as a semigroup of functions (usually called mappings in this context) on a set. The proof is nearly identical with that of the traditional Cayley theorem in that the representation arises through the action of each element a of the given semigroup S on S itself—the only complication is that we need to adjoin an identity element to S if it does not already possess one in order to ensure that the representation is faithful. The resulting monoid is denoted by S^1, where 1 denotes the identity element.

The other ingredient in the connection to recognizable sets is the free monoid A^* over a given alphabet A. This is none other than the set of all possible finite strings or words formed from A. The operation is *concatenation* meaning that if u and v are two such strings then their product in A^* is simply the string uv. The empty string ε is also permitted and this acts as the identity element of A^* making it a monoid. If we delete the empty string we are left with A^+, the free semigroup on A.

The idea of a homomorphism (or sometimes just *morphism*) is a universal algebraic idea and applies equally well to semigroups and monoids in the same way as it does to groups: a morphism $\alpha : S \rightarrow T$ between semigroups is a mapping that respects products in that $\alpha(ab) = \alpha(a)\alpha(b)$. (In the case of monoids we need to insist that a morphism maps the identity to the identity as that does not follow automatically otherwise.) The free semigroup and monoid are then free in the usual sense. They are the freest algebras containing the given generating set A and any semigroup (resp. monoid) generated by A is a homomorphic image of A^+ (resp. A^*).

A subset L of A^*, invariably referred to as a *language* in this context, is said to be *recognized* by a semigroup S if there is a morphism $\alpha : A^* \rightarrow S$ such that $L = \alpha^{-1}(P)$ for some subset P of S. In words, a language is recognizable (by a semigroup) if it is the inverse image of a subset of S under a morphism from the free monoid.

The algebraic and automata based versions of recognizablity are equivalent: a language can be recognized by an automaton if and only if it can be recognized by a finite semigroup. The proof in each direction is quite short but a little technical and so will not be recounted here (see for example *Finite Automata* by M. V. Lawson). However, in outline it runs as follows. If L is recognized by a finite semigroup S as above, we consider the automaton that has S^1 as its set of states, initial state 1, and the action of each letter a mimics that of $\alpha(a)$, multiplying elements of S (on the left if we compose mappings from right to left); the automaton's set of accepting states is deemed to be P, where $L = \alpha^{-1}(P)$, whence, by construction, the language accepted by this automaton is L.

The converse direction goes by way of the *transition* monoid of an automaton A. Each letter a of A acts on the states of A to give a mapping in the full transformation semigroup whose base set is the state set of A. This in turn induces a morphism α from A^* into this full transformation semigroup S. The language L of the automaton is then recognized by S and α as $L = \alpha^{-1}(P)$, where P is the set of all words of the form $\alpha(u)$, where u is a word of L.

Once the basics have been established in this manner we enjoy the full freedom to study recognizablity through either of these two equivalent approaches, these being the machine or the algebraic viewpoint. Some results are more transparent when seen algebraically. For example, if L is recognized by S then the language of reversed words, L^r, is recognized by the left–right dual semigroup S^r whose multiplication $\circ$ is defined by $a \circ b = ba$ (where the latter product is in S), using the same morphism and subset of S that were the ingredients in the recognition of L.

Showing that certain languages are not recognizable, such as the languages of all palindromes or the languages of all words with equal numbers of a's and b's, is particularly simple. Roughly speaking, arguments that appeal to the so-called Pumping Lemma in automata theory are replaced by arguments involving idempotents in the semigroup approach (a finite semigroup always has at least one *idempotent*, that is a member a that equals its own square). There is a natural interplay between features in one theory and the other. For example, the transition monoid of the minimal automaton of a language is a common morphic image of every semigroup that recognizes that language. A semigroup that recognizes a language also recognizes all its so-called quotients (which are therefore recognizable too). More broadly, a natural class of semigroups, known as a *variety*, corresponds to what is known as a *stream* or *variety* of languages, which is a class of languages that is closed under various natural operations, which include the boolean set operations and quotients.

These theories are very much two sides of the one coin. However, one side of this coin is comparatively neglected!

Page 133 Least common multiple of the first ten counting numbers

In general, to find the lcm n of a given set of numbers we write n as a product of powers of primes. The power p^r required for each prime number p is the greatest power of p that is a factor of any of the numbers in the set. For the integers $1, 2, \ldots, 10$ the only relevant primes are $2, 3, 5,$ and 7. The highest power of 2 involved is $2^3 = 8$, while we have $3^2 = 9$, and the primes 5 and 7 only ever arise as single powers. Therefore our least common multiple in this case is $2^3 \times 3^2 \times 5^1 \times 7^1 = 8 \times 9 \times 5 \times 7 = 2{,}520$.

Page 136 Finite lattices have meets and joins of arbitrary sets

Let L be our finite lattice and suppose that S is any non-empty set of nodes of L with $|S| = n \geq 1$. We show by induction on n that the meet of S exists: the proof for joins is exactly the same with the symbol $\vee$ replacing $\wedge$ throughout. There is no trouble anchoring the induction, for the claim is clearly true in the cases where $n = 1$ or $n = 2$ so let us assume that $n \geq 3$. Take a node $u \in S$ and consider the set $S' = S \setminus \{u\}$. Since $|S'| = n - 1$ it follows by induction that the meet of S' exists and we denote it by v. We then claim that $u \wedge v$ is the meet of S.

By definition of the $\wedge$ operation, $u \wedge v \leq u$ and $u \wedge v \leq v \leq s$, where $s \in S'$. Hence $u \wedge v$ is a lower bound for the set S. Let w denote an arbitrary lower bound of S. Then $w \leq s$ for any $s \in S'$ and since v is the greatest lower bound of S' we infer that $w \leq v$. Moreover, since w is a lower bound for S, in particular $w \leq u$. Hence w is a common lower bound of u and v and therefore $w \leq u \wedge v$. This proves that $u \wedge v$ is indeed the greatest of the common lower bounds of all the members of S.

In particular, taking S to be the full vertex set of L, we see that any finite lattice has a minimum node and a maximum node.

Chapter 7

Page 139 Prim's Algorithm

We show by a rather subtle induction that Prim's algorithm, where a tree is extended in a greedy manner until it spans the network, does always yield a minimal spanning tree.

We suppose that the network N has n nodes and let the trees constructed via Prim be listed as $T_1, T_2, \ldots, T_{n-1}$, where T_i has as its edges the list $(e_1, e_2, \ldots, e_i)$. Let T be a minimal spanning tree of N that has as many edges in common with T_{n-1} as possible. We demonstrate the result by showing that T and T_{n-1} are in fact identical.

Suppose to the contrary that $T \neq T_{n-1}$ and let $e_j = (a,b)$ be the first edge chosen by Prim that is not in T. Let P be the path in T from a to b (since T is a spanning tree for N, there is a unique path within T between any two nodes of N). Next let e^* be an edge of P between a node in T_{j-1} and a node not in T_{j-1} (since a is a node of T_{j-1} and b is not, such an edge e^* exists). Now the weight of e^* is at least as great as that of e_j, for otherwise e^* would have been chosen by Prim ahead of e_j when forming the tree T_j. We then delete e^* from T and replace it by e_j to form a new network T' of weight no more than T that has one more edge in common with T_{n-1} than does T. However, by construction, T' is still connected (this needs a moment's thought) and has n nodes and $n - 1$ edges so is indeed a spanning tree for N, contradicting our original choice of T. Hence we must drop the assumption that $T \neq T_{n-1}$ and so Prim's algorithm will always produce a spanning tree of N of minimal weight.

Page 140 **Counting spanning trees**

The *incidence* matrix of a network is a binary $n \times n$ matrix where the (i,j)th entry is a 1 if node i is adjacent to node j and is 0 otherwise. (This presupposes that we have numbered the nodes in some (perhaps arbitary) order. Coding a network as a matrix is a natural way to store it in a computer from which point particular algorithms of interest may be carried out.)

The *Kirchhoff matrix* of a simple network is obtained by taking the adjacency matrix, swapping all the 1's for -1's and replacing each diagonal 0 by the degree of the corresponding node. Its significance lies in the theorem that all the cofactors of any such matrix are identical and their common value equals the number of spanning trees of the network.

For example, if we take our network N to be a square with one diagonal (so that it is one edge short of being K_4) the Kirchhoff matrix of N would be:

$$\begin{pmatrix} 3 & -1 & -1 & -1 \\ -1 & 2 & 0 & -1 \\ -1 & 0 & 2 & -1 \\ -1 & -1 & -1 & 3 \end{pmatrix}$$

Kirchhoff's Theorem says that the number of spanning trees of a connected simple network is equal to any of the cofactors of the matrix. In this example, if we use the $(1,1)$ cofactor we obtain the answer as:

$$2 \begin{vmatrix} 2 & -1 \\ -1 & 3 \end{vmatrix} - \begin{vmatrix} 0 & -1 \\ 2 & -1 \end{vmatrix} = 2(6-1) - (0-(-2)) = 10 - 2 = 8,$$

so there are eight spanning trees to be found (six paths of length 3, while two trees have nodes of degree 3).

It is instructive to work out an example like this in the case of a tree, for you will know if you have carried out the calculation correctly as the answer must be 1! The calculation of such cofactors for a network of more than half a dozen nodes soon becomes too laborious to carry out by hand. However, the theorem does provide a convenient approach for a computer to use.

What is more, Kirchhoff's Theorem can also be used to verify a theoretical counting result that goes right back to Cayley: the number of *labelled trees* on n nodes is n^{n-2}. By a *labelled network* in this context we mean each of the nodes has its own distinct label, usually a number. The key observation is that every labelled tree on n nodes is a spanning tree for the complete labelled network K_n and conversely, any spanning tree on K_n is a labelled tree on n nodes. The problem then comes down to applying Kirchhoff to K_n. (Although not an induction argument, it is worth checking the first few cases: for example, for $n = 3$ there are $3^1 = 3$ labelled trees, each being determined by the label of the node of degree two.)

We need, therefore, to calculate any of the cofactors of the matrix $M = M_n$ all of whose diagonal elements are $n - 1$ and whose off diagonal elements are -1.

The $(1,1)$-cofactor of M is the determinant of a matrix that is almost M_{n-1} except that the diagonal entries are still $n - 1$. The trick is to add every row to the first, and then the first row to each of the others (which leaves the value of the determinant invariant). The matrix that results has the first row consisting entirely of 1's, the first column entirely of 0's (except for the initial 1) and the remainder of the matrix is nI_{n-2}, where I denotes the identity matrix. Since this matrix is upper triangular, its determinant is simply the product of the diagonal elements, which evidently is n^{n-2}. This is just one of many clever proofs of Cayley's enumeration of labelled trees.

Page 144 Designing the one-way system

What remains to be proved is that if the network N has no bridges, then N is strongly connected, and in order to do this we show by induction on k that it is always possible to drive from 1 to k and back again in the one-way system that the algorithm provides.

The $k = 1$ case is clear so we take $k \geq 2$ and suppose that the claim holds for all lesser values. The node labelled k was assigned its number by virtue of being adjacent to some other node l that was already labelled. In that case $l < k$ and by induction it is possible to drive from 1 to l and from l to 1 within the system. By the way the tree is constructed and labelled, the arc between l and k is directed $l \to k$, and so there is a directed path from 1 to k via l. However, we need to find a way of also getting back from k to 1.

Suppose all the nodes adjacent to k in the undirected network have been labelled prior to k (so that the algorithm, if it continues, will backtrack from k). Now, there must be some node adjacent to k apart from l as otherwise the edge lk would be a bridge. Take one such node with label $m < k$. Since the edge mk is not part of the spanning tree (otherwise k would already have been labelled) its orientation is $k \to m$ and by induction there is a directed path from m to 1, thus yielding a directed path from k to 1.

However, there is still the case where the algorithm does not backtrack after labelling k but rather proceeds on to new nodes labelled $k+1, k+2, \ldots, k+t$ say, where $t \geq 1$ and $k+t$ is a node from which the algorithm does backtrack. If any of these nodes happens to be adjacent to a node labelled $m < k$ we obtain a directed path from k to 1 via that node. If, to the contrary, all the nodes $k+i$ $(1 \leq i \leq t)$ were only adjacent to other nodes of this same kind, there would be no path from k to l that did not use the edge lk, and once again, kl would be a bridge, giving the required contradiction to complete the proof.

Page 145 **Acyclic digraphs have linear orderings on their nodes**

Suppose that N is an acyclic (not necessarily connected) digraph with n nodes. Begin at any node u and take a directed walk for as long as you can. As there are no directed cycles, eventually we meet a node v that has no arc leading out of it. Delete this node and its incident arcs from N to give a digraph N' on one fewer nodes that is still acyclic. By induction, we can order the nodes $v_1, v_2, \ldots, v_{n-1}$ in such a way that there is an arc from v_i to v_j only if $i < j$. Putting

$v_n = v$ then gives a linear ordering of all the nodes of N with the same property.

Chapter 8

Page 161 **Maximum flow equals minimum cut**

As has already been pointed out, the maximum flow through the network N cannot exceed the capacity of the minimum cut, so it is enough to show there is a cut whose capacity equals that of an attainable flow. A *flow* θ assigns a non-negative integer to each arc a in a specified direction so that the flow does not exceed the capacity of any arc and the flow into a node (apart from the source and sink) equals the outflow from that node. An arc is called *saturated* if the flow through it equals its (maximum) capacity. The assumption that the flow through an arc is integral is harmless—by scaling to suitably small units this can be seen not to act as a genuine restriction but merely allows us to bring to bear arguments based on discrete units.

Let θ be a maximum flow in N with s and z respectively denoting the source and sink of N, so that the flow begins at the source s and ends at z. We introduce two disjoint sets of nodes V and W as follows: let G denote the underlying network; put a node u in V if there is some path in G: $s = v_0 \to v_1 \to v_2 \to \ldots \to v_{m-1} \to v_m = u$, with the property that each edge $(v_i v_{i+1})$ represents either an unsaturated arc (one not carrying its maximum capacity) or there is a positive backflow in the direction $(v_{i+1} v_i)$. The set W is merely the complementary set to V. By definition, V is not empty as it at least contains the source of the flow, s.

We show that z lies in W. If not, then z lies in V and so there is a path $s = v_0 \to \ldots \to z$ of the type described. We now perturb the flow by a small positive amount ε. We choose ε so as not to exceed the amount needed to saturate any arc of the first type and not to exceed the flow through any arc of the second type (backflow arcs). We now increase the flow through arcs of the first type by ε and decrease the flow through the backflow arcs by ε. (This is possible as it will not cause any violation of the inflow = outflow condition at each node.) The net effect of this is to increase the overall flow to the value of $\theta + \varepsilon$, which is impossible as θ represents a maximum flow. Therefore it is indeed the case that z lies in W and not V.

The question is now settled by considering the set A of all arcs $a = (x,y)$ from V to W. This collection is a cutset as s is in V and z is in W, so that all flow passes out of V into W. Moreover, every arc (x,y) in A must be saturated for otherwise, since x lies in V, then y would also if (x,y) were not used up to maximum capacity. It follows that the capacity of the cut A is, as claimed, the same as the maximum flow in the network.

Page 165 Bipartite = all cycles have even length

The fact that $K_{3,3}$ has all cycles of length at least 4 was used in Chapter 4 to show that it was not planar. In a bipartite network based on two disjoint sets of vertices, G and B, any cycle passes between these two sets a certain number, let us say k times, and so the cycle has even length, $2k$.

Conversely, let N be a network in which all cycle lengths are even and since the following argument can be applied one component at a time, there is no harm in taking N to be connected. By the *distance* between any two nodes we mean the length of a shortest path joining them.

Begin with any node u and let G be the set of all nodes, including u, whose distance from u is even and let B consist of all nodes that are not in G. Then B and G form the two 'parts' of the bipartite network and to justify this claim we need only check that no two nodes in G are adjacent, and similarly for B.

To see this, suppose that two nodes, u_1 and u_2 in G were adjacent. Consider paths P_1 and P_2 of shortest possible lengths from u to u_1 and u_2 respectively. Let x be the final common point of these paths from u to u_1 and u_2. The length of the initial segments of P_1 and P_2 from u to x must be equal, for if the first were longer than the other, say, then P_1 could not be an initial segment of a shortest path from u to u_1. Let us then call this common length k, and let l_1 and l_2 denote the distance from x to u_1 and u_2 respectively. Then, if k is even, then so are l_1 and l_2, while if k is odd, then so are l_1 and l_2. In either case, by taking the terminal segments of P_1 and P_2 together with the edge $u_1 u_2$, we now obtain a cycle of length $l_1 + l_2 + 1$, which is odd, contrary to the assumption that N has no cycles of odd length. A similar contradiction arises if we assume that two nodes in B are adjacent, and this completes the proof.

Page 167 **Hall's Marriage Lemma**

Here is a direct induction argument for the tricky direction of the lemma. That is, every set of k girls collectively is prepared to marry at least k of the boys for all $1 \le k \le n$ and we need to show that there will then necessarily be a matching for all the girls. The argument here, due to Halmos and Vaughan, is a short but subtle induction on n, the number of girls. There is no trouble of course if there is only one girl so let us take $n \ge 2$.

First suppose that it were the case that for all k with $1 \le k \le n - 1$, every set of k girls had a set of suitable boys of size at least $k + 1$. In this case, marry off one suitable pair, leaving us with $n - 1$ girls, and it is still the case that any k of them ($1 \le k \le n - 1$) is suited to at least k boys. We can then marry off the remaining $n - 1$ girls by induction.

The alternative is that the supposition is false, in which case there is some set of k girls ($1 \le k \le n - 1$) whose set of suitors numbers exactly k as well. Since $k < n$ we can, by induction, marry off these k girls to their k boys, leaving a set of $n - k$ girls remaining. We can now marry off these $n - k$ girls by induction, *providing Hall's condition holds for the remaining girls and boys*, and this has to be checked.

But it does! Any h say of the remaining girls ($1 \le h \le n - k$) does have a set of suitors of size at least h, for suppose that this condition were somehow violated for some particular set of girls of size h, so that they are suitable for fewer than h of the remaining boys. Consider then this set of h girls together with the original set of k girls considered earlier. Then this collection of $k + h$ girls is only suited to a set of fewer than $k + h$ boys. (This is because none of the k girls is suited to any of the remaining boys—the key point of the argument.) This, however, violates the original condition on subsets of girls, so represents a contradiction. Therefore, the remaining $n - k$ girls can be married by induction and the proof is complete.

There are many other proofs of Hall's Lemma, which is one of the most useful tools in combinatorics. Although often short, every proof has a tricky bit that you have to think about!

Page 169 Menger, Marriage, and Maximum Flows

We assume that the condition of the Marriage Lemma holds and we begin with the same augmented network as was used in the demonstration of the Marriage Lemma from the Max Flow Theorem, and indeed the argument is much the same.

Given that there are n girls in G, a complete matching for the girls is equivalent to a set of n node disjoint paths from the source to the sink. By the nodal version of Menger therefore, it is sufficient to show that any nodal cutset S for the pair of nodes s and z has at least n nodes. Let X and Y be the respective subsets of G and of B that together form the set S. If we assume that X has k nodes, which is less than n, then, by the Hall condition, the remaining $n - k$ girls are collectively linked to at least $n - k$ boys. Each of the nodes representing these boys must be in Y, as otherwise there would be a path from s to z that did not use any nodes from the nodal cutset S. Therefore the size of S must be at least $k + (n - k) = n$.

Although Menger's Theorem takes some work, it is relatively easy to show that the edge and node versions of the theorem are equivalent, meaning that if you believe one of them, then you must believe the other as well, as each can be deduced from the other.

The Max Flow Theorem can also be inferred from Menger's Theorem by the device of replacing an arc with a capacity of k units by k multiple edges between the same nodes. A flow now corresponds to a set of edge-disjoint paths between the source and sink. By Menger's Theorem, the maximum value for this is the minimum size of any edge cutset, which in turn corresponds to the minimum capacity of a cutset. Conversely, we can deduce Menger's Theorem from the Max Flow Theorem by giving each edge capacity of one unit. Applying the Max Flow Theorem then gives the required conclusion that the maximum size of a set of edge-disjoint paths in the network equals the minimum size of an edge cutset.

Chapter 9

Page 181 **Insanity Cubes**

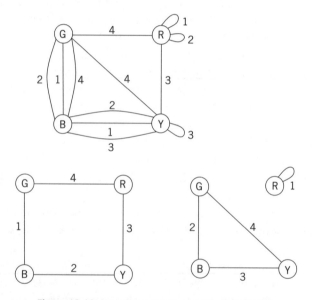

Figure 10.10 Insanity network and labelled factors

In this instance the network of the puzzle is as in Figure 10.10 with two suitable labelled factors as shown. Unlike the puzzle in the text, the second factor necessarily has two components; we follow the labelling of the 3-cycle in a cyclically consistent fashion (say clockwise) to obtain a consistent colouring of the back and front of the tower with the three colours blue, green, and yellow: cube 1 carries the colour red both front and back.

Page 183 **Two litres of wine**

Again, we can specify the state of play with an ordered pair (x,y) where x and y are the measures of the contents of the 7- and the 4-litre jugs respectively, bearing in mind that $0 \leq x + y \leq 10$, as 10 is the capacity of the big jug. The minimum sequence of pourings in this case turns out to be

$$(0,0) \to (0,4) \to (6,4) \to (6,0) \to (2,4)$$

and so in just four steps we have a 2-litre portion in the 7-litre jug (with the 4-litre and 10-litre jugs both left holding 4 litres each).

Page 195 **Reconstructing the RNA chain**

    G-fragments: *AACUG, UAG, A, AG, AG, AG, G*

    U, C-fragments: *U, AGAAC, AGAGA, GGAGU*

There are two abnormal fragments, *A* and *AGAGA*, and so the longer forms the end of the chain. The interior extended bases and unsplittable fragments are

    interior extended bases: *U, AG, G, AG*

    unsplittable fragments: *A, AG, AG, AG, G, U*

The two unsplittable fragments that are not extended bases are *AG* and *A*, and since the chain terminates with *AGAGA*, it begins with *AG*. The digraph of the problem is then given by the final diagram.

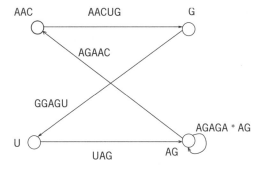

Figure 10.11 Digraph for RNA chain

There is only one Euler circuit that ends with the starred arc (which is in this case a loop), giving the arc labels

AGAAC AACUG GGAGU UAG AGAGA*AG

and so the reconstituted RNA chain is

AGAACUGGAGUAGAGA

REFERENCES

Chapter 1

My source for the counting facts on number of trees is Frank Harary's classic text *Graph Theory* published by Addison-Wesley (1972), and the figure on chemical isomers is after Figure 1.4 there.

Chapter 2

The circular puzzles are taken from *The Official Book of Circular Sudoku* by Caroline and Peter Higgins, Plume Press, New York (2006). The mastermind example is based on Exercise 2.1.1 of Alan Tuckers's *Applied Combinatorics*, Wiley & Sons, New York (1980).

Chapter 3

I have relied largely on the book by Albert-Laszlo Barabasi, *Linked: How Everything is Connected to Everything Else*, Perseus Books (2002), for the account given here of the small-world phenomenon and scale-free networks.

The quote of Aristotle's view of friendship comes from the account of that philosopher in Bertrand Russell's timeless classic *History of Western Philosophy* now published in the UK by Routledge (2004).

Chapter 4

The three problems in the Rabbits out of Hats section (Guarding the Gallery, the Sylvester–Gallai Problem, and Sperner's Lemma) follow the proofs given in *Proofs from the Book* (Springer, 1999) by Martin Aigner and Gunter Ziegler.

Chapter 5

The description of the Chinese Postman Problem follows that of chapter 10 of *Discrete Mathematics with Graph Theory*, Prentice-Hall (1998), by Edgar Goodaire and Michael Parmenter.

Chapter 6

The Mealy machine (b) is that of examples 6, chapter 5 of *Theory of Computation, An Introduction*, Jones & Bartlett (1996), by James L. Hein.

Chapter 9

The Instant Insanity examples in the text come from exercise 2 of chapter 11 of Tucker's book (see Chapter 2 above). The example of formation of a Huffman code follows that of exercise 12.17 in *Discrete and Combinatorial Mathematics*, 4th edn., Addison-Wesley (1999), by Ralph P. Grimaldi. The worked example of reassembling of RNA chains is exercise 10.3 1(e) of *Discrete Mathematics and Graph Theory* (see Chapter 5 above) and the additional example is problem 1 in section 10.3 of the same text.

Chapter 10

The argument for the existence of Ramsey numbers is adapted from that in *Ramsey Theory*, Wiley & Sons (1980), by Ronald Graham et al. For Hamiltonian cycles see Robin Wilson, *Introduction to Graph Theory*, Oliver & Boyd (1972).

FURTHER READING

As mentioned in the text, in mathematics the general area that is the subject of this book is known as *Graph Theory*, especially among the mathematical fraternity. It may seem an odd name as almost everyone knows that a graph usually means a plot of one measurement against another, typically a quantity such as sales, inflation, or velocity as a function of time. A classic mathematical text is *Graph Theory* by Frank Harary and it was the source of some of the counting facts about the number of trees on a given number of nodes and the isomers of saturated hydrocarbons. Overall, though, this is a pure mathematics text and is not much interested in applications to operational research. Robin Wilson's little book *Introduction to Graph Theory* gives a rapid outline of the subject along with the classical applications of matching theory.

I normally would hesitate to recommend a mathematics text book as a source for someone who had developed a passing interest in the subject but *Discrete Mathematics*, as this general area is known, is a more accessible part of the subject. There are dozens of good texts on discrete mathematics, often known also as *Finite Mathematics*, and the titles often include the terms graph theory, computation, or more mysteriously *combinatorics*. This last word is nothing to be afraid of but it does sound esoteric. It is a word not found in most dictionaries, so a general reader might take it as a signal that the book is not for them. To explain, combinatorics is that part of mathematics to do with counting, which may also sound rather strange, for don't we all know how to count already? Examples will best clarify: questions such as how many different noughts and crosses games are there? how many genuinely different sudoku puzzles are there? and how large does a network have to be before a certain kind of clique is bound to arise? are examples of combinatorial questions. Sometimes precise answers can be given, although on tougher questions we often have to be satisfied with bounding the answer between two extremes and talking of how fast the figure in question increases as various parameters vary. This can be very important when deciding whether or not a problem will become too unwieldy to handle if it is allowed to grow more complicated.

Of the many books you will come across of this type, I like *Discrete Mathematics with Graph Theory* by Goodaire and Parmenter. This was my source particularly for the description of the Chinese Postman Problem and the problems on the recovery of RNA chains. Ralph Grimaldi's *Discrete and Combinatorial Mathematics: an Applied Introduction* goes further in the direction of coding theory and algebraic applications. Both of these books are weighty tomes in the modern American style. I also like *Applied Combinatorics* by Alan Tucker, which has more the feel of an old-fashioned mathematics book. The account of Instant Insanity and some other novelties was due to this text, which you might find is more for a serious mathematics student. It does, however, explain how to go about solving some important but messy problems like finding Hamiltonian cycles in sizeable networks.

The source for *Circular Sudoku* is the recent book by myself and my daughter Caroline: *The Official Book of Circular Sudoku* (Plume Press). That book explains the puzzles, including variants that do not feature here, and how to solve them. Techniques for solving ordinary sudoku will be supplied by any decent search engine on the Web.

The book by Albert-Laszlo Barabasi, *Linked: How Everything is Connected to Everything Else* has certainly been influential and has increased the general interest in networks of all kinds. The description of the nature of the internet, which was pioneered by Barabasi and his colleagues, is summarized in the second chapter of this book. Paul Hoffman's biography on Erdos, *The Man Who Loved Only Numbers*, Hyperion, New York (1998) is an interesting account of the life of a mathematician. Some reviewers fear that it slips into caricature of the man and the subject in places but the general public will find there some real insight into what mathematicians get up to.

The Theory of Computation, an Introduction, Jones & Bartlett (1996) by James Hein is better than some of the big hardbacks on discrete mathematics as, in addition to lots of algorithms and examples, there is more of the underlying mathematics as well. It also deals well with all forms of automata. For more on *Automata and Languages* I would recommend either the book of that title by John M. Howie and also *Finite Automata* by Mark V. Lawson, although be warned, these are also mathematics books.

The trickiest mathematics described in this book came from the celebrated *Proofs from the Book* by Martin Aigner and Gunter Ziegler and included the Guarding the Gallery, and the Sylvester–Gallai and Brouwer theorems. These are relatively difficult but serve to show how subtle even quite 'elementary' mathematics can be. You have to be wary when a mathematician talks of an 'elementary' proof. The word in this context means only that the proof does not call upon sophisticated techniques, such as the use of complex variables. An 'elementary' proof can be difficult to follow and even more difficult to discover! The title, 'Proofs from the Book', is a reference to the Platonic book, imagined by Paul Erdos, which was a home for all of the very best proofs. In Heaven, those brave souls who have proved themselves worthy, will be free to read *The Book* to their hearts' content!

INDEX

accepting state (of automaton) 120
Ackermann's function 153
adjacent (nodes, edges) 7
AKS primality test 156
Alcuin of York 184
algorithm 104
  Bellman–Ford 149
  complexity of 151–3
  Dijkstra 148–50
  Fleury 103, 220
  Kruskal's 139
  Nearest Neighbour 145–6
  Prim's 139, 226
  shortest path, *see* Dijkstra
alphabet 120
antipodes 86
antisymmetric 134
Appel, Kenneth 73
arc 1, 112
  saturated 230
Ariadne's thread 186
automaton 119–32
  non-deterministic 127
  pushdown 128

backtrack 142, 187
Barabasi, Albert-Laszlo 39
barycentre 218
base (biological) 191
  extended 193
  interior 193
base (of induction) 197
binary string 189–91
Boolean Satisfiability problem 157
bound:
  greatest lower 132
  least upper 132

lower 132
upper 132
bridge 103, 140–1, 228–9
Brouwer, L. E. J. 91–3
  fixed point theorem 90–2, 217–19
bubble sort 151

Cayley, Arthur 9, 228
  theorem 223
Chain Rule 9
checkers 20
chemical isomers 9–10
Chinese Postman Problem 105–10
Chomsky, Noam 129
Chomsky heirarchy 129
cipher 188
  RSA 188
circuit 7
  directed 110, 114, 117–19, 194–5
  Euler 53, 101
  Hamilton 145
  simple 7
clique 59, 212
code 188–91
  Huffman 189–91
  prefix 188
colouring 63–5
combinatorics 60
  of nodes 81
compactness 218–19
complement 126
complete enzyme digest 193
complexity, *see* algorithm
concatenation (or words) 223
Connect Four 19
context-free languages 128
context-sensitive languages 129

convex shape 81
Cook, Stephen 157
counterfeit coin problem 22,
  201
Cretan maze 185–6
critical path 150
cutset 160–2
  nodal 170
cycle 7
  directed 112, 144
  Hamilton 53–6, 177–9, 221

de Bruijn graph 114–19
  sequence 115–19
De Morgan, Augustus 64
degree, *see* node
digraph 111
  Hamiltonian 112
  semi-Hamiltonian 112
  strongly connected 112
Dijkstra, Edsger 148
dodecahedron 55
Durer, Albrecht 24

edge 1
  endpoint 199
  incident with 7
  multiple 7, 45
enzyme 191
Epimenides of Knossus 14, 93
Erdös, Paul 38–9
  number 38
Euclid's Algorithm 133
Euler, Leonhard 2, 26, 43–8, 101–5
  circuit 101, 219–20
  equation 75–7
  path 101
exponential decay 207–8

factor of a network 180
  edge disjoint 180
factor of a word 122
factorial 202
five-colouring a map 73, 214–16
fixed point theorems 90–3
flow 166, 230
Floyd–Warshall algorithm 150
four–colour map problem 63–74

fragment:
  abnormal 192
  normal 194
  unsplittable 193

Go 20
Graeco-Latin square 25–7
graph 8, 111
  bipartite 164, 231
  directed 111
great circle 86
greedy algorithm 139
Guthrie, Francis 63

Harem Problem 170–1
Haken, Wolfgang 73
Hall's Condition 164, 172,
Hall's Marriage Lemma 163–7,
  232–3
Hamilton, William Rowan 53, 55
Hand-Shaking Lemma 48–50, 99
Heawood P. J. 73
Higgins, Caroline 31
highest common factor 133
homomorphism 126, 223
hub 37–40

icosahedron 55
incident with 7
in-degree 112
induction 82, 198
input 120
Instant Insanity 175–81, 234
intermediate growth 153
internet 2
isomorphism 69
isthmus 103

join 132, 225–6

$K_{2,2,2}$ 66–7
$K_{3,3}$ 52
K_5 52
Kant, Emmanuel 43
Karinthy, Frigyes 36
Kempe, A. P. 73
Kirchhoff, Gustav 9, 43, 139
  Theorem 227

Kleene, Stephen 126
   Theorem 126
Knaves and Knights 11, 200
Königsberg, Bridges of 43–5
Kuratowski, K. 77

label:
   permanent 148
   temporary 148
labryrinth, *see* maze
language:
   of automaton 120
   product of 125
   quotient 127
   rational, *see* set, regular
   theory 120
   variety of 225
Latin rectangle 30, 172–3
Latin square 23–31, 172–3
   normalized 202
   number of 202
   orthogonal 202
lattice 132–6
   concept 136
least common multiple 133,
   225
linear algebra 140
logic puzzles 10–15, 17–23
log linear complexity 152
log-log paper 209
loop 8, 62
Lo-shu 23

machine:
   deterministic 155
   non-deterministic 127, 155
   Turing 129
magic square 23–4
Marriage Problem 163
Mastermind 31–2
matching 172
matroid 4
matrix:
   incidence 139
   Kirchhoff 227
Max Flow Min Cut Theorem 161–3,
   230–1, 233
Maximum Flow Problem 161

maze 185–8
   centre of 187
   of the minatour 186
Mealy machine 129–32
meet 132, 225–6
Mei-ko Kwan 105
Menger, Karl 169
   Theorem 169, 233
   nodal form 170
merge sort 152
Mersenne, Marin 39
minimal criminal 53
monoid 223
   transition 224
Moore machine 129
mutiliated chessboard 96

n-colourable 81
National Football League
   49
net (of a solid) 176
network (net) 1
   acyclic 144
   bow-tie 54
   bipartite 164, 172, 231–2
   colouring 81
   complementary 57
   complete 52, 71
   component of 7
   connected 101
   contractible 80
   cubic 55
   de Bruijn, *see* graph
   directed 3, 111–14
   dual 65
   face of 74
   homeomorphic 78
   isomorphic 69
   monchromatic 210
   Petersen 79–80
   planar 51, 66, 74–80
   plane 67, 70
   Platonic 55, 79
   regular 79
   scale-free 41, 208
   simple 8, 62
   small world, 208
   trivalent 79

node 1
  adjacent 7
  balanced 114
  cutset 170
  degree of 6–7
  even 8
  incomparable 134
  isolated 7
  odd 8
  valency of 9
noughts and crosses 17–18
NP problem 155
NP-complete 156
number:
  chromatic 81, 214
  irrational 93
  rational 92

octahedron 55
Othello 21
out-degree 112

P problem 155
palindrome 123
paradox 93
  Russell's 95
parity argument 96
path 7
  directed 112
  edge disjoint 169
  Euler 53, 101–3, 106
  node disjoint 169
  successful 12
perfect matching 172
perfect ranking 113
PERT 150
Pigeonhole Principle 59–61, 124,
  213–14
power law 40, 207
Pumping Lemma 125

Ramsey, F. P. 59
  number 59, 209
  problem 57
Ranks and Regiments problem
  27
Reversi 21
RNA 191–5, 235–6

search:
  breadth-first 183
  depth-first 142, 183
  tree 11
set:
  independent 212
  intersection of 125
  product of 125
  regular 125, 222–5
  theory 95
  union of 125
Shortest Path Problem 109, 148–50
sink 112
source 112
Sperner's Lemma 96–9, 218
state (of automaton) 120
  accepting 120
  initial 120
  terminal 120
subset sum problem 153–6
successor function 130
Sudoku 27–9, 171, 203
  circular 30–1, 203–5
Sylvester–Gallai Theorem 86–90

target 206
tetrahedon 55
tic-tac-toe 17–18
time:
  exponential 153
  polynomial 152
torus 85
tournament 111–13
  semi-Hamiltonian 221
trail 8
traffic 140–5
  one-way 228–9
transitive 134
Travelling Salesman Problem (TSP) 55,
  145–7, 156
tree 5–9, 197–200
  binary 155
  chess 19
  directed 200
  family 34
  full complete binary 154
  height of 201
  Huffman 190

labelled 228
leaves of 201
m-ary 201
root of 11, 200
spanning 137–40, 227–8
ternary 23, 201
triangulation 82, 97–9, 216–18
mesh of 217
truth value 94
two tribes 10–15

Van der Waerden's Theorem
213
vertex (-ices) 1

walk 8
weight 55, 147
of a tree 138
word 120
factor of 122
World Wide Web 3